This book is due for return not later than the
last date stamped below, unless recalled sooner.

# Myelopeptides

# Myelopeptides

**R. V. Petrov**
**A. A. Mikhailova**
**L. A. Fonina**
*Russian Academy of Sciences*

**R. N. Stepanenko**
*Russian Ministry of Health*

**World Scientific**
*Singapore • New Jersey • London • Hong Kong*

*Published by*

World Scientific Publishing Co. Pte. Ltd.

P O Box 128, Farrer Road, Singapore 912805

*USA office:* Suite 1B, 1060 Main Street, River Edge, NJ 07661

*UK office:* 57 Shelton Street, Covent Garden, London WC2H 9HE

**British Library Cataloguing-in-Publication Data**
A catalogue record for this book is available from the British Library.

**MYELOPEPTIDES**

ISBN 981-02-3507-0

Printed in Singapore.

## Preface

Myelopeptides are a group of regulatory bone marrow peptides that we have isolated and studied. After deciphering their primary structure, we synthesized these peptides. Our first papers, in which the ability of bone marrow cells to stimulate the antibody genesis is reported, were published in 1969. In 1975, this effect was shown to be mediated by a soluble factor that is produced in short term cultures of bone marrow cells. We named it the SAP — that is, the stimulator of antibody production.

The solution to the question on its chemical nature was complicated by the fact that its acting base is present in minor fractions of the supernatant of bone marrow cell culture. The first results that confirm SAP heterogeneity and its peptide nature were obtained in 1978. This was the reason for us to re-name the SAP and introduce the term "myelopeptides" into the literature.

The principle of our work has always been "from the activity to the substrate". In other words, we did not progress from the isolation of all possible peptides to the subsequent screening of all possible activities in this peptide pool. We rather searched for the basis of the physiological activity that was found in biological experimental models *in vitro* or *in vivo*. Considering the small quantities responsible, this approach was not easy but it proved to be fruitful and yielded peptides with the known activities. The myelopeptides isolated and studied were named MP-1, MP-2 etc. The data collected in this book present materials on revelation and study of the following peptides:

- ◆ MP-1 (Phe–Leu–Gly–Phe–Pro–Thr) — a peptide with immunocorrecting activity;
- ◆ MP-2 (Leu–Val–Val–Tyr–Pro–Trp) — a peptide that displays antitumor effect;
- ◆ MP-3 (Leu–Val–Cys–Tyr–Pro–Gln) — a peptide that stimulates macrophage phagocytosis;

- MP-4 (Phe–Arg–Pro–Arg–Ile–Met–Thr–Pro) — a cell-differentiating factor;
- MP-5 (Val–Val–Tyr–Pro–Asp) and MP-6 (Val–Asp–Pro–Pro). The study of their biological activities is in progress now.

Our search for homologous sequences through the protein–peptide bank of PIR showed that the amino acid sequences of MP-1 and MP-2 are identical to those of conservative fragments of α-chain (33–38) and β-chain (31–36) of vertebrate hemoglobin, while peptides MP-3, MP-4, MP-5, and MP-6 have no homologous sequences with proteins deposited in the PIR bank.

Along with experimental data outlined in the book, we also present some materials on the clinical application of a medicinal preparation, Myelopid, consisting of a native complex of myelopeptides that are produced by porcine bone marrow cells. Here, we summarize some results on its clinical efficacy in the therapy of patients with secondary immunodeficient states that are provoked by severe surgical traumas, by some lung diseases, and by osteomyelitis. Myelopid is effective in the treatment of some forms of hypogammaglobulinemia, leukemia or immunodeficiency that develops after chemo- or X-ray therapy.

The deciphering of the structure, function and mechanism of action of the isolated individual myelopeptides provides some promising prospectives in order to develop medicinal preparations of a new generation — preparations that produce a natural directed correcting action and that have no side effects.

Our experimental and clinical investigations are currently in progress.

*Rem Petrov*
*June 1998*

## List of Abbreviations

| | |
|---|---|
| AFC | antibody forming cells |
| Con A | concanavallin A |
| Cy | cyclophosphamide |
| FITC | fluorescein isothiocyanate |
| HGG | horse gammaglobulin |
| HL-60 CM | conditioned medium from HL-60 human leukemia cell line |
| HPLC | high performance liquid chromatography |
| IL-1 | interleukin-1 |
| IL-2 | interleukin 2 |
| LPS | lipopolysaccharide |
| MI | maturation inductor |
| MP | myelopeptide |
| PHA | phytohemagglutinin |
| PMA | phorbol myristate acetate |
| PWM | pokeweed mitogen |
| RhITC | rhodamine isothiocyanate |
| SRBC | sheep red blood cells |
| TDF | T cell differentiating factor |

# Contents

# Part 1
# New Immunoregulatory Bone Marrow Peptides, Myelopeptides

## 1. Discovery of Bone Marrow Factor(s)

### 1.1. Cell Cooperation at the Level of Mature Antibody Producers

The late 1960s to early 1970s were marked by outstanding discoveries in cellular immunology. I. Roitt postulated the existence of two cell types of bone marrow and lymphoid origin and named them B and T lymphocytes, respectively (Roitt et al., 1969). J.F.A.P. Miller and G.F. Mitchell showed in experiments that T–B lymphocyte interaction is necessary to induce antibody genesis (Miller and Mitchell, 1968; Mitchell and Miller, 1969). The first papers demonstrating the necessity of three-cell types' interaction to realize the immune response — B, T lymphocytes and macrophages — were published in 1969 (Petrov, 1969; Roitt et al., 1969; Berenbaum, 1969). Having only information on the antigen, B lymphocytes cannot transform into mature antibody-forming cells (AFC) without additional signals produced by macrophages and T lymphocytes. The macrophage, whose role was previously assumed to be limited to the engulfment of foreign substances only, turned out to initiate the specific immune response as the cell which phagocytoses the antigen, processes it to an immunogenic form and then presents it to B and T lymphocytes.

The three-cell cooperation scheme of induction of the immune response provided a new insight into the mechanisms underlying the functioning of the immune system as well as a better understanding of its complicated multicomponent nature. It became clear that there is a complex network of interrelating information signals which provide good coordination of the functioning of all interacting immunocompetent cells. Such signals were soon revealed. These are numerous cytokines (for example, interleukins, interferons, colony-stimulating factors) that could be figuratively named *molecular couriers of immunity.*

At first, such complicated cell–cell interactions were only considered necessary for the initiation of the immune reaction to induce B lymphocyte differentiation to mature plasma cells which synthesize and secrete specific antibodies. It seemed that the induced process of antibody formation does not need any additional signals to transform lymphocytes into antibody-producing plasma cells and proceeded only according to a pre-determined program. However, we managed to reveal another level of immunocompetent cell interaction. It appeared that cell cooperation takes place not only in the inductive, but also in the productive phase of the immune response. This was first displayed in experiments on the joint cultivation of lymphoid and hemopoietic cells obtained from immunized and non-immunized animals (Petrov and Mikhailova, 1969, 1972).

Lymph node cells obtained from mice on the fourth day after their secondary immunization with horse gamma globulin (HGG) were cultivated together with syngeneic bone marrow cells from non-immunized mice. After 20 h cultivation, the synthesis of antibodies, nonspecific immunoglobulins (Ig) and other water-soluble proteins was measured by means of $^{14}$C-glycine incorporation. Antibodies and nonspecific Ig were successively isolated from the conditioned medium using specific immunosorbents. After removing all Ig, other water-soluble proteins were precipitated by trichloracetic acid.

It was shown that antibody synthesis in the mixed cultures increased two- to threefold as compared to that in the monoculture. Nonspecific Ig synthesis increased by only 1.5 times. The synthesis of water-soluble proteins of non-Ig nature did not change (Table 1).

Table 1. Stimulation of synthesis of antibodies, nonspecific Ig and
other water-soluble proteins under joint cultivation of lymph node cells
from immunized mice and syngeneic intact bone marrow cells.

| Cultured cells | Number of experiments | Protein synthesis, cpm | | |
|---|---|---|---|---|
| | | Antibodies | Nonspecific Ig | Other water-soluble proteins |
| Immune lymph node cells | 7 | 810±241 | 1340±310 | 9670±1840 |
| Bone marrow cells | 7 | No synthesis | 915±113 | 32920±8832 |
| Immune lymph node cells + bone marrow cells | 7 | 2389±635 | 3640±410 | 48560±6948 |
| Stimulation coefficient | | 2.95 | 1.61 | 1.14 |

Such selective stimulation of antibody synthesis in the mixed culture points to processes taking place which are directed to the enhancement of antibody formation, and not at the intensification of total protein synthesis.

Table 2. Increase in AFC number under joint cultivation
of immune lymph node and intact bone marrow cells.

| Antigen | Cell types and genotype of the cell donors | | Number of experiments | AFC/$10^6$/nucleated cells | | Stimulation coefficient | P |
|---|---|---|---|---|---|---|---|
| | Immune lymph node cells | Intact bone marrow cells | | Mono-culture | Mixed culture | | |
| HGG | A | A | 6 | 70±26 | 176±21 | 2.5 | ·0.05 |
| SRBC | CBA | CBA | 5 | 140±4 | 300±9 | 2.1 | ·0.01 |
| | A | A | 6 | 100±9 | 430±27 | 4.3 | ·0.01 |

The determination of AFC number in the mono- and mixed cultures showed that the joint cultivation of immune lymph node and intact bone marrow cells for 20 h resulted in a two- to threefold increase in AFC amount. This effect was observed for both, the soluble (HGG) and the corpuscular (sheep red blood cells, SRBC) antigens (Table 2).

The similar stimulation coefficients evaluated by AFC number and the level of antibody synthesis show that the effect of increased antibody formation in the mixed culture is not caused by increased intensity of total protein synthesis by antibody producers but is related to the increase in the number of cells producing antibodies with the same specificity.

It was established that the effect of stimulation of antibody formation in the mixed culture reaches its maximum when lymph node cells from mice at the peak of the secondary immune response were used, that is, when the number of mature differentiated plasmatic cells is maximal.

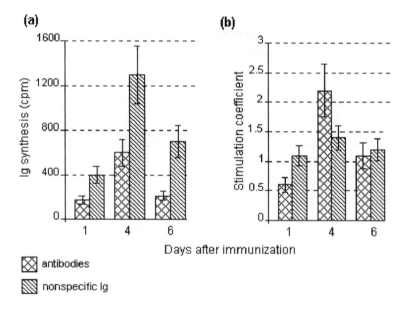

Fig. 1. Intensity of antibody and nonspecific Ig synthesis in the immune lymph node cell culture (a) and stimulation of the synthesis of these proteins in the mixed culture (b) as a function of time after the immunization of cell donors.

Figure 1a demonstrates the intensity of antibody and nonspecific Ig synthesis in the immune lymph node cell culture depending on the period after the immunization of cell donors with HGG.

The maximal synthesis of these proteins in the immune lymph node cells was observed on the fourth day after the immunization. The stimulation coefficients of Ig synthesis in the mixed culture were maximal when the lymph node cells were obtained from mice during this period (Fig. 1b). On the first and sixth days, when Ig synthesis in lymph node cells was minimal (Fig. 1a), the antibody stimulation effect of bone marrow cells was insignificant or totally absent (Fig. 1b).

These experiments indicate that stimulation of antibody formation in the mixed culture can only manifest itself when a large number of mature antibody producers are present in the immune cell population. It points out the ability of the bone marrow to influence the intensity of antibody formation at the peak of the immune reaction, provided that an AFC clone is present.

It is well known that the immune response to the antigen stimulus composes several stages that depend on each other but are regulated by different mechanisms. In the inductive phase of antibody genesis after antigen recognition, active proliferation and differentiation of precursor cells take place, whereas in the productive phase of the immune response, synthesis and secretion of antibodies occurs in mature antibody producers.

Our experiments have shown that the increase in the AFC number in the mixed culture of immune lymph node and intact bone marrow cells is not accompanied by enhancement of cell proliferation. It suggests the presence of so-called *reserve* or *silent* cells in the population of mature antibody producers which can be recruited in the process of antibody formation under the influence of a signal produced by bone marrow cells. In distinction to cell interactions initiating the immune response, the revealed phenomenon was termed *cell cooperation at the level of mature antibody producers* (Petrov and Mikhailova, 1972).

We suppose that the induction of immunogenesis and the subsequent development of immune response are provided by two different types of cell interaction. The "Macrophage–T lymphocyte–B lymphocyte" cooperation includes immunopoietic differentiation of precursor cells and causes their multiplication and maturation of AFC (the first level of cooperation). At the next stage of the immune response with a significant amount of mature cells involved in antibody production, cell cooperation at the level of mature antibody producers is developing. Additional cells, which are ready for protein synthesis but produce no antibodies, are involved into antibody formation (the second level of cooperation). This hypothetical scheme is in conformity with the experimental data. According to the scheme, every antibody producer can recruit two other cells for antibody formation. Under these circumstances the actual yield of antibodies synthesized increases two-to threefold as compared to the predicted one.

This cell interaction type originally revealed *in vitro* was later reproduced in our experiments *in vivo*. We used a modified model that is usually employed to study transplantation immunity reactions in regional lymph nodes after local administration of allogeneic cells.

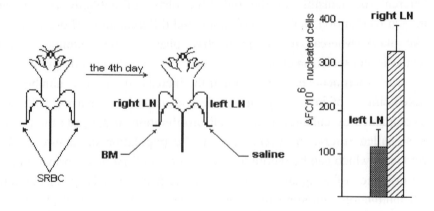

Fig. 2. Assaying the stimulatory effect of bone marrow cells on antibody production *in vivo*. LN, lymph node cells; BM, bone marrow cells.

Syngeneic bone marrow or lymph node cells were injected into one hind footpad of mice ($1.5 \times 10^7$ cells per mouse) on the fourth day of the secondary immune response to SRBC. Physiologic saline was injected into the contralateral footpad as a control. 24 h after the injection, the number of AFC was evaluated in both popliteal lymph nodes. The scheme of the experiment and the results are shown in Figure 2.

Table 3 represents stimulation coefficients of antibody and nonspecific Ig synthesis in CBA mouse immune lymph node cells *in vitro* after the addition of bone marrow cells obtained from mice of A strain, rats, rabbits, calves or syngeneic mouse cells. One can see from the table that stimulation of antibody formation is observed during the cultivation of cells obtained from different animal species. The stimulation coefficients of Ig synthesis in xenogeneic and allogeneic mixtures practically did not differ from those in syngeneic mixtures.

Table 3. Stimulation of antibody and nonspecific Ig synthesis in the mixed cultures of mouse immune lymph node cells with syngeneic, allogeneic or xenogeneic bone marrow cells.

| Components of the mixture | | Number of experiments | Stimulation coefficient | | | |
|---|---|---|---|---|---|---|
| Immune lymph node cells | Intact bone marrow cells | | Antibodies | P | Nonspecific Ig | P |
| CBA mice | CBA mice | 3 | 2.9±0.25 | <0.05 | 1.57±0.07 | <0.05 |
| CBA mice | A mice | 3 | 2.4±0.16 | <0.05 | 1.5±0.03 | <0.05 |
| CBA mice | Rats | 3 | 1.98±0.38 | <0.05 | 1.36±0.04 | <0.05 |
| CBA mice | Rabbits | 4 | 1.80±0.5 | <0.05 | 1.5±0.01 | <0.05 |
| CBA mice | Calves | 3 | 2.2±0.31 | <0.05 | 1.6±0.02 | <0.05 |

The absence of histocompatibility barrier in the antibody stimulating effect produced by bone marrow cells opens a perspective to use the revealed phenomenon in practice to achieve immunostimulation in the late phases of antibody genesis. Cell cooperation at the level of mature antibody producers is apparently one of the links in immunoregulatory processes providing normal development of immune reactions, and the bone marrow plays a significant role in the regulation of antibody formation.

## 1.2. Stimulator of Antibody Producers, SAP

The revealing of cell cooperation at the level of mature antibody producers entailed a comprehensive study of this cell interaction type, in the first place, it was necessary to clarify if some soluble factors mediate these intercellular interactions. Experiments on separation of interacting cells by means of a Millipore membrane showed the participation of humoral bone marrow factor(s) in the stimulation of antibody formation at the peak of immune response.

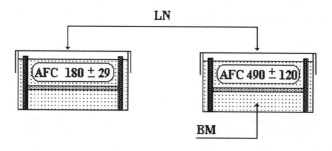

Fig. 3. Increase in AFC number under joint cultivation of immune lymph node and intact bone marrow cells separated by a Millipore membrane; LN, lymph node cells; BM, bone marrow cells.

A two-chamber vessel with a Millipore membrane (diameter of pores 25 nm) impenetrable for cells was used in a 20 h cell cultivation. The upper chamber contained lymph node cells obtained from mice at the peak of the secondary immune response to SRBC, that is mature AFC; the lower chamber contained bone marrow cells from non-immunized mice or culture medium. Figure 3 represents the results of a standard experiment. The presence of bone marrow cells under the membrane caused a 2.5-fold increase in the AFC number in the population of immune lymph node cells as compared to the control vessel with the culture medium in the lower chamber. It was concluded that bone marrow cells produce a factor(s) that penetrates through the membrane and causes an increase in the number of antibody producers in the population of mature AFC.

The presence of humoral bone marrow factor(s) enhancing antibody formation in the productive phase of the immune response is also confirmed by the data on stimulating activity of the supernatant obtained after cultivation of bone marrow cells. When a conditioned medium from a 20 h bone marrow cell culture was added to lymph node cells obtained from mice at the peak of the secondary immune response to SRBC, the stimulation effect on antibody formation could be compared to that obtained after addition of intact bone marrow cells (Table 4).

Table 4. Changes in AFC number in the population of immune lymph node cells under the influence of intact bone marrow cells or conditioned medium.

| Cultures | $AFC/10^6$ nucleated cells | Stimulation coefficient | P |
|---|---|---|---|
| Immune lymph node cells (control) | 149±31 | | |
| Immune lymph node cells + bone marrow cells | 420±42 | 2.9 | <0.05 |
| Immune lymph node cells + conditioned medium | 371±45 | 2.5 | <0.05 |

The stimulating activity of the humoral bone marrow factor(s) is displayed not only *in vitro*, but also at the level of a whole body. In a series of experiments, the conditioned medium from bone marrow cell culture was injected into one hind footpad of mice on the fourth day of the secondary immune response to SRBC. Physiologic saline was injected into the contra-lateral footpad which served as a control. After 24 h, the AFC number was determined in the popliteal lymph nodes. The results showed that the injection of supernatant obtained after cultivation of bone marrow cells like the injection of bone marrow cells causes an approximate threefold increase of AFC number in the regional lymph node as compared to the control (Table 5).

Table 5. Increase in AFC number in the regional lymph node in the productive phase of the immune response under local injection of bone marrow cells or conditioned medium obtained after their cultivation.

| Injected agents | Number of experiments | AFC/$10^6$ nucleated cells | Stimulation coefficient |
|---|---|---|---|
| Bone marrow cells Physiologic saline | 4 | 145±30 50±20 | 2.8 |
| Conditioned medium Physiologic saline | 4 | 129±18 43±9 | 3 |

So, a humoral factor(s) produced by bone marrow cells, that is capable of increasing the amount of AFC at the peak of immune response, was revealed. This factor was termed *stimulator of antibody producers*, SAP (Petrov *et al.*, 1975).

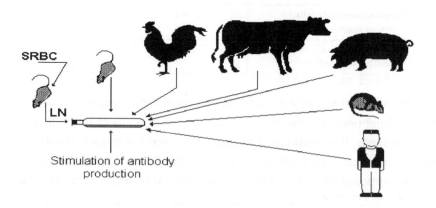

Fig. 4. Increase in AFC number in the mouse immune lymph node cell culture after the addition of supernatants from bone marrow cell cultures of various animal species and man. LN, lymph node cells.

SAP is produced by bone marrow cells of various animal species and man and does not have species specificity. The addition of supernatants obtained after cultivation of mouse, hen, calf, pig, rat or human bone marrow cells to lymph node cells obtained from mice at the peak of the secondary immune response to SRBC caused a 1.5- to twofold stimulation of antibody formation. The scheme of these experiments is presented in Figure 4.Further experiments on the study of the nature and functional activity of SAP were conducted with the use of supernatants of 20 h pig bone marrow cell culture. SAP was the first among the discovered bone marrow factors influencing the intensity of antibody formation. At around the same time, Canadian (Duwe and Singhal, 1976) and Russian (Petrov *et al.*, 1976, 1977) researchers demonstrated a suppressive effect of bone marrow cells on antibody genesis. If SAP stimulated antibody formation at the peak of the immune response, that is at the level of mature antibody producers, then the suppressive action of bone marrow cells on the immune response is displayed at the inductive phase of antibody genesis, that is at the level of precursor cells. The analysis of all these data suggests the participation of bone marrow humoral factors in regulatory processes controlling the development of antibody genesis.

## 1.3. SAP Structural Heterogeneity

Accumulation of SAP in the supernatant of the bone marrow cell culture raised a possibility of isolating an active fraction containing this mediator. We performed successive chromatography of the supernatant of a 20 h pig bone marrow cell culture on Sephadexes with various pore sizes (G-100, G-50, G-25 and G-15). The antibody-stimulating activity of the fractions obtained was assessed in an *in vitro* test schematically represented in Figure 5.

The mice were immunized with SRBC. On the fourth day after reimmunization, we obtained lymph node cells from these mice and cultivated them with or without the tested fraction. On the next day, we determined the AFC number in the test and control cultures. It appeared that antibody-stimulating activity was revealed in the fraction which was eluted in the region of markers having molecular mass of 0.5–3 kD (Petrov *et al.*, 1986; Mikhailova and Petrov, 1987). Figure 6 shows a chromatogram of the active material on Sephadex G-15.

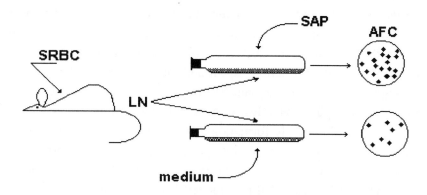

Fig. 5. Assaying SAP biological activity *in vitro*. LN, lymph node cells.

Proceeding from the absorption profile at λ=206 nm, we grouped six peaks in four fractions. The first of them was eluted as a free volume, the rest as an internal one. Fractions 1 and 2 displayed antibody-stimulating activity. Since fraction 2 contained 80% of the whole material, the substances responsible for antibody-stimulating effect were apparently contained mainly in fraction 2 and had a molecular mass of around 1.5 kD. Electrophoresis of fraction 2 demonstrated heterogeneity of the active substance. Electrophoresis was performed on plates with a small cellulose coating in the buffer. Ninhydrin or fluorescamin staining were used for detection. The results of the staining with both reagents were similar. Figure 7 shows that fraction 2 obtained from Sephadex G-15 is divided into five further zones.

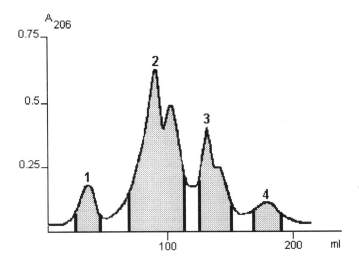

Fig. 6. Gel chromatography of SAP-containing fraction ( $\lambda$= 206 nm).

Only zones 4 and 5 displayed antibody-stimulating effect. To elucidate whether the active substances are possible components of the culture medium, we compared electrophoregrams of fraction 2 and preparation of the culture medium obtained in the same manner. It appeared that active zones 4 and 5 were not present in the medium preparation.

We also conducted a comparative separation of equivalent quantities of fraction 2 and a model mixture of four amino acids which were contained in this fraction along with other amino acids. One can see from Fig. 7 that electrophoretical mobility of active zones 4 and 5 and that of amino acids in the model mixture differed significantly. So we can conclude that at least in active zones 4 and 5 these amino acids are in a bound state.

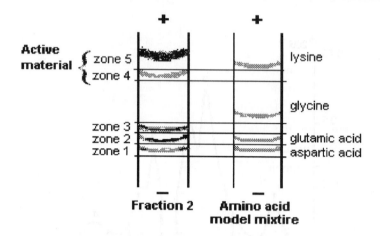

Fig. 7. Electrophoresis of fraction 2 and model amino acid mixture (schematic representation).

The data presented suggest that the bone marrow mediator SAP is heterogeneous in its composition and is a mixture of biologically active substances produced by the bone marrow cells. This work was a starting point to conduct a comprehensive study of the nature and functions of bone marrow immunoregulatory mediators.

## 2. Myelopeptides: A Group of Bone Marrow Bioregulatory Mediators

The bone marrow, being one of the central organs of the immune system, plays a very important role both in immuno- and hemopoiesis. It is the source of stem cells that give rise to all hemopoiesis pathways, including lymphopoiesis. In the bone marrow, pre-T and pre-B lymphocytes are generated; their further maturation in the thymus or in the bone marrow then governs the development and functioning of T and B systems of immunity in mammals. Therefore it is quite evident that the humoral factors produced by bone marrow cells can play a substantial role in the functioning of the immune system.

Being a hemopoietic organ, the bone marrow provides differentiation and maturation of various blood cells that suggests the production of several differentiation factors by bone marrow cells.

According to the current knowledge on cytokines, the immune system mediators are regarded as a part of a complex chain of molecular signals that provide for concerted functioning of various systems in the organism. In this connection, the data on the polyfunctionality of cytokines and their production in both nonlymphoid and lymphoid organs — in the central nervous system, in particular — are of interest. All these point to their general biological significance. It is possible that the bone marrow mediators not only participate in the regulation of immuno- and hemopoiesis, but that they also provide some intersystem interactions. That is why our study of humoral bone marrow factors was headed for search of bone marrow substances with various bioregulatory properties.

We investigated the nature and bioregulatory effects of a fraction with antibody-stimulating activity (SAP) isolated from the supernatant of porcine bone marrow cell culture. We tested it in several experimental models that revealed various biological properties (Petrov and Mikhailova, 1987; Mikhailova and Petrov, 1992).

## 2.1. Peptide Nature of Bone Marrow Mediators

It is known that substances of a peptide nature play an important role in bioregulation. Peptides and polypeptides of endogenous origin participate in many bioregulatory processes. Binding with specific receptors on the membrane of the target cell, they "switch on" its function essential in that moment because the expression of the receptors on the appropriate cell is controlled by definite biological reactions. Bone marrow mediators, which we studied, obviously participate in these complex molecular events. To determine their chemical nature, we at first used some indirect methods (Petrov *et al.*, 1980).

The treatment of the active fraction (obtained on Sephadex G-15) with some proteolytic enzymes, followed by evaluation of its antibody-stimulating activity, allowed us to determine to which group of bioactive molecules the revealed bone marrow mediators belong. Trypsin, papain, pronase and RNAase were used in these experiments. It is known that trypsin is a highly specific enzyme that breaks the polypeptide chain at the level of carboxyl arginine and lysine groups. Papain has a lower specificity to the substrate, catalyzing the hydrolysis of not only peptides but also that of amines, esters and thioesters. Papain displays the highest specificity towards cysteine containing peptides. Finally, pronase (bacterial protease) is characterized by the lowest degree of specificity hydrolyzing 80% of all peptide bounds in large proteins. Ribonuclease attacks the –P–O– bounds in nucleotides independent of the nature of their heterocyclic bases.

The results of our experiments on enzyme treatment of the active fraction are presented in Table 6. The hydrolysis by RNAase had no marked effect on the biological activity of the fraction. The proteolysis effect was displayed according to the substrate specificity of the enzymes used. Treatment with trypsin did not affect the antibody-stimulating effect of the fraction while treatment with papain slightly decreased the effect. However treatment with pronase significantly decreased the antibody-stimulating effect of the fraction. This means that molecules responsible for stimulation of antibody formation are probably composed of some peptides.

Table 6. Changes in antibody-stimulating effect of fraction 2
on the immune lymph node cells after enzyme treatment.

| Cultivated cells | Number of experiments | Number of AFC/10⁶ nucleated cells | Stimulation coefficient |
|---|---|---|---|
| Immune lymph node cells (control) | 8 | $68\pm32$ | |
| Immune lymph node cells + fraction 2 | 8 | $164\pm34$ | 2.4 |
| Immune lymph node cells + trypsin-treated fraction 2 | 8 | $143\pm25$ | 2.09 |
| Immune lymph node cells + papain-treated fraction 2 | 8 | $116\pm17$ | 1.7 |
| Immune lymph node cells + pronase-treated fraction 2 | 8 | $101\pm16$ | 1.4 |
| Immune lymph node cells + RNAase-treated fraction 2 | 8 | $143\pm26$ | 2.1 |

The peptide nature of bone marrow mediators stimulating antibody genesis is also substantiated by the results of inhibitor analysis. Cycloheximide was used to inhibit protein synthesis, RNA synthesis was blocked by actinomycin D, while DNA synthesis was blocked by mitomycin C. The bone marrow cells were cultivated for three hours at 37°C with various doses of inhibitors, then they were washed and added to the lymph node cells from immunized mice, jointly cultivated for 18-20 h and the effect of stimulation of antibody formation was assessed in the mixed cultures. The efficiency of the inhibitors was measured by means of $^{14}$C-glycine incorporation into proteins, $^{14}$C-uridine into RNA and $^{3}$H-thymidine into DNA.

The data presented in Table 7 show that protein and RNA synthesis is essential for the production of antibody-stimulating substances by bone marrow cells, while DNA synthesis is not necessary.

Table 7. Stimulation of antibody formation in the mixed culture of immune lymph node and intact bone marrow cells treated with inhibitors of protein, RNA or DNA synthesis.

| Inhibitor | Number of experiments | Concentration, μg/ml | Synthesis inhibition (%) | | | Stimulation coefficient |
|---|---|---|---|---|---|---|
| | | | protein | RNA | DNA | |
| Cycloheximide | 3 | – | – | – | – | 2.52±0.16 |
| | 3 | 25 | 82.4±2.0 | | | 1.22±0.11 |
| | 3 | 500 | 97.2±0.8 | | | 1.02±0.02 |
| Actinomycin D | 3 | – | – | – | | 2.53±0.30 |
| | 3 | 1 | 18.0±1.5 | 23.9±2.8 | | 1.13±0.09 |
| | 3 | 4 | 41.0±3.2 | 60.0±1.2 | | 0.93±0.13 |
| Mitomycin C | 3 | – | – | – | – | 2.52±0.16 |
| | 3 | 33 | – | – | 91±1.8 | 2.51±0.21 |

All these data suggest that bone marrow cells produce biologically active peptides that can mediate the regulatory functions of the bone marrow with respect to the immune system. These peptides were termed *Myelopeptides* (MPs).

## 2.2. Biological Activities of MPs

As shown above, MPs were revealed due to their ability to stimulate antibody production. The data obtained do not exclude the fact that the fraction isolated from the supernatant of porcine bone marrow cell culture consists of several biologically active substances which possess several other activities along with the antibody production stimulating one.

In this connection, we have conducted a search for biological activities of MPs by testing the isolated fraction in various experimental models revealing different biological effects: immunostimulating, immuno-corrective, cell differentiating, as well as neurotropic ones. It allowed us to confirm the fact that the analyzed fraction consists of compounds responsible for a spectrum of biological effects and are obviously connected with bone marrow functions.

## 2.2.1. Immunostimulating Effects of MPs

The immunostimulating activity of MPs was first analyzed by the ability of the fraction isolated on Sephadex G-15 to increase antibody production at the peak of the immune response *in vitro*. The addition of MPs to the lymph node cells obtained from mice on the fourth day of the secondary immune response to SRBC resulted in a two- to threefold increase in the AFC number after 18-20 h cultivation. Other stimulators, like thymus peptides and mitogens (PHA, Con A, LPS), did not display stimulating effect under these experimental conditions (Table 8).

Table 8. Stimulating effect of MPs on AFC number in the culture of immune lymph node cells obtained from mice at the fourth day after the secondary immune response to SRBC.

| Stimulating agent | Dose, μg/ml | Number of experiments | AFC/$10^6$ nucleated cells | Stimulation coefficient | P |
|---|---|---|---|---|---|
| – | – | 8 | 113±13 | – | – |
| Bone marrow cells | 1:1 | 8 | 411±42 | 3.64 | <0.001 |
| MPs | 100 | 8 | 183±27 | 1.61 | <0.05 |
| | 50 | 8 | 259±26 | 2.30 | <0.001 |
| | 25 | 8 | 245±13 | 2.16 | <0.001 |
| Thymus peptides | 10 | 8 | 128±12 | 1.13 | >0.05 |
| PHA | 10 | 8 | 115±13 | 1.02 | >0.05 |
| Con A | 10 | 8 | 124±114 | 1.10 | >0.05 |
| LPS | 1 | 8 | 136±15 | 1.20 | >0.05 |

The stimulation of antibody formation at the peak of the immune response under the addition of MPs was reproduced with various antigens.

Figure 8 represents the stimulation coefficients of AFC number in the spleens or antibody titers in the serum of animals immunized with various antigens. The figure shows that MPs stimulate by two to five times the antibody formation to the antigens used, and the effect of stimulation of

antibody formation is more pronounced when antigens with weak
immunogenicity are used.

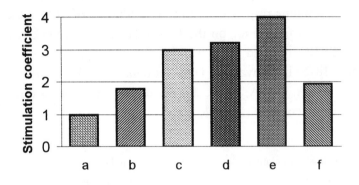

Fig. 8. MPs-induced stimulation of antibody production in response to various
antigens: (a) control; (b) SRBC; (c) microbial antigens; (d) HGG; (e) flaviviruses; (f) Thy-1 antigen.

Let us illustrate the influence of MPs on the production of antiviral
antibodies. Mice were infected with clone 3 of Japanese encephalitis or with
Langat TP-21 virus of tick-borne encephalitis. The viruses were injected
intraperitoneally at a dose of $10^5 LD_{50}$ ($10^5$-fold greater than the dose causing
a 50% animal death). On the tenth day after virus injection, we determined
the level of antiviral antibodies in the serum of the animals in a virus
neutralization reaction as well as in a hemagglutination inhibition reaction.
On the third day, some of the infected animals were intravenously injected
with MPs at a dose of 50 μg/mouse. The rest of the animals were injected
with MPs at the same dose thrice on the third, fifth and eighth days after the
infection. The control animals were injected with physiologic saline. The
antibodies were determined in the serum pools of animals in various groups.
The data presented in Table 9 demonstrate that a single injection of MPs
resulted in a four- to eightfold increase of antiviral antibody titers in the

virus neutralization reaction, and in a fourfold increase in the haemagglutination inhibition reaction.

Table 9. Effect of MPs on the production of antiviral antibodies
in mice infected with flaviviruses.

| Animal groups | Titers of antiviral antibodies | | | |
| --- | --- | --- | --- | --- |
| | C3 (JE) | | TP-21 (Langat) | |
| | Virus neutralization reaction | Hemagglutination inhibition reaction | Virus neutralization reaction | Hemagglutination inhibition reaction |
| Single MPs injection | 1:80 | 1:320 | 1:40 | 1:160 |
| Triple MPs injection | 1:80 | 1:160 | 1:160 | 1:320 |
| Control | 1:10 | 1:80 | 1:10 | 1:40 |

A triple injection of MPs resulted in an eight- to 16-fold increase of antibody titers (antigen neutralization reaction), and in a two- to eightfold increase of haemagglutinins. A similar effect of MPs was registered when the ability of the serum obtained from the same animal groups to neutralize the cytopathogenic effect of the virus was assessed (Petrov *et al.*, 1983-a).

Injection of MPs also positively influenced the survival of the infected animals. The triple injection of MPs into mice infected with the virus of clone 3 of Japanese encephalitis resulted in a twofold increse of their survival.

As already mentioned, we supposed that the increase in antibody formation under the influence of bone marrow mediators occurs at the expense of recruiting additional "silent" cells into the process of antibody synthesis. These "silent" cells are ready for protein synthesis but do not produce antibodies. The presence of a "silent" population among mature antibody producers at the peak of the immune response is confirmed by experiments in some other model systems.

Thus, the possibility to recruit "silent" cells at the peak of the immune response into antibody synthesis without cell-fission was shown in the experiments on evaluation of the role of cell proliferation in the increase of the AFC number in the culture of spleen cells obtained from immunized mice. The suppression of cell proliferation by means of colchicine or mitomycin C resulted in the inhibition of AFC production in suspensions of spleen cells obtained from mice at early stages of the immune response. At the same time, the addition of the inhibitors to the spleen cells obtained from mice at the peak of immune response did not prevent the enhancement of AFC number.

We managed to develop an experimental model which enabled us to concentrate the "silent" cells involved in the antibody formation under the influence of MPs (Petrov *et al.*, 1981). In order to do so, we allowed the lymph node cells obtained from immunized mice to run through the column with Sephadex G-10 that adsorbs most of antibody-secreting cells. The cell population that had passed through the column contained only 17-30% AFC (Table 10).

Table 10. Effect of bone marrow cells or MPs on AFC number
in a population of mouse immune lymph node cells
before and after their fractionation on a Sephadex G-10 column.

| Stimulating agent | Number of experiments | AFC/$10^6$ nucleated cells | | | | Stimulation coefficient | |
|---|---|---|---|---|---|---|---|
| | | Before fractionation | | After fractionation | | Before frac- tionation | After frac- tionation |
| | | Control | Experiment | Control | Experiment | | |
| Bone marrow cells | 6 | 86±19.2 | 218±28.8 | 15±3.5 | 208±28.6 | 2.53±0.17 | 13.9±3.38 |
| MPs | 3 | 133±9.4 | 248±28.0 | 44±9.0 | 166±11.0 | 1.75±0.18 | 6.0±0.75 |

When intact bone marrow cells or MPs-containing fraction were added to this AFC impoverished population, the stimulation of antibody formation was more pronounced as compared to their addition to a non-fractionated cell culture. The stimulation coefficients increased from 2.53 to 13.9 when bone marrow cells were used, and from 1.75 to 6.0 when an MPs-containing fraction was added (Table 10).

The blockade of antibody synthesis in "silent" cells could be explained by an inhibitory effect of T suppressors. In this case, MPs initiate antibody synthesis in the "silent" cells thus abolishing the suppressive activity of T lymphocytes. We checked this suggestion experimentally (Mikhailova et al., 1976).

T suppressors were obtained according to the method of M. Taussig (Taussig, 1974). The lethally irradiated mice (CBA×C57Bl)F$_1$ were intravenously injected with $10^8$ syngeneic thymocytes together with $2\times10^8$ SRBC. On the seventh day after the injection, spleen cells obtained from these mice were incubated for 6-7 h with SRBC in vitro. The cells were washed and further used in the experiments as T suppressors.

T suppressors were added to lymph node cells obtained from mice at the peak of the secondary immune response to SRBC or HGG and incubated for 15 h. Then the AFC number or antibody synthesis intensity was determined.

The results of these experiments showed that T suppressors inhibited antibody formation in an immune lymph node cell population by five times. It did not depend on the type of antigen used — specific (SRBC) or nonspecific (HGG) ones (Table 11).

Table 11. Effect of T suppressors on antibody formation
in the population of immune lymph node cells.

| Cultivated cells | AFC/$10^6$ nucleated cells under SRBC immunization | Stimulation coefficient | P | Antibody synthesis under HGG immunization, counts/100 sec | Stimulation coefficient | P |
|---|---|---|---|---|---|---|
| Immune lymph node cells | 580±11 | | | 5001±473 | | |
| Immune lymph node cells + T suppressors | 120±29 | 0.21 | <0.05 | 1491±361 | 0.30 | <0.01 |

When MPs were added to the culture along with T suppressors, one could not observe any inhibitory effect of T lymphocytes. MPs produced a stimulating effect on antibody formation like that in the absence of T suppressors (Table 12).

Table 12. Effect of MPs on AFC number in the population of immune lymph node cells under the addition of T suppressors.

| Cultivated cells | Number of experiments | AFC/$10^6$ nucleated cells | Stimulation coefficient |
|---|---|---|---|
| Immune lymph node cells | 4 | 580±110 | |
| Immune lymph node cells + MPs | 4 | 1830±430 | 3.16 |
| Immune lymph node cells + T suppressors + MPs | 4 | 2080±490 | 3.59 |

The possibility that MPs compete with the suppressive activity of T lymphocytes was also demonstrated when T suppressors were induced by means of a mitogen (Con A).

Con A was injected at a dose of 25 µg/mouse 2 h after secondary immunization with SRBC thus ensuring the appearance of T suppressors in the spleens of these mice (Rich and Pierce, 1973). The AFC number in the spleens of such animals was reduced to 50% of that of the control group (Table 13).

Table 13. Effect of MPs on the Con A-induced suppression of antibody production.

| Added agent | Number of experiments | AFC/$10^6$ nucleated cells | Stimulation coefficient |
|---|---|---|---|
| Control | 4 | 888±20.8 | |
| Con A | 4 | 460±12.8 | 0.51 |
| Con A+ MPs | 4 | 1223±17.1 | 1.38 |

When MPs together with Con A were inoculated into mice at the peak of the secondary immune response to SRBC, the suppressive activity of Con A-induced T lymphocytes was not detected and antibody formation increased (Table 13).

A series of experiments with anti Lyt-2 monoclonal antibodies confirmed the ability of MPs to block the suppressive activity of T lymphocytes. Lymph node cells obtained from mice at the peak of the secondary immune response to SRBC were treated with anti Lyt-2 serum. The removal of Lyt-2+ cells (conventional T suppressors) evoked an increase in AFC number in the population of immune lymph node cells that could be compared to the stimulating effect of MPs (Fig. 9b, c). The addition of MPs to this culture practically did not influence the level of antibody production as compared to the culture depleted of T suppressors (Fig. 9d).

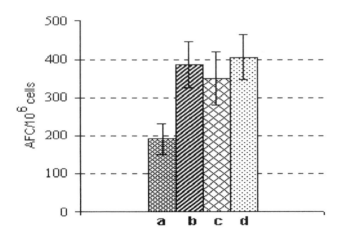

Fig. 9. Effect of anti Lyt-2 monoclonal antibodies and MPs on the AFC number in the population of immune lymph node cells obtained from SRBC-immunized mice. (a) Immune lymph node cells (control); (b) immune lymph node cells + MPs; (c) immune lymph node cells treated with anti Lyt-2 serum; (d) immune lymph node cells treated with anti Lyt-2 serum + MPs.

The antibody-stimulating effect of MPs is obviously connected with their ability to abolish the suppressive activity of T lymphocytes.

The immunostimulating activity of MPs is displayed not only by an increase in antibody production at the peak of the immune response. It was also revealed that MPs can influence immunity by enhancing macrophage phagocytosis and stimulating the macrophage lysosomal enzymes, thus evoking a more rapid removal of pathogenic microorganisms from the bodies of infected animals (Petrov *et al.*, 1988).

Peritoneal macrophage phagocytosis was determined in mice infected with *S. typhimurium* 415. MPs were injected intraperitoneally 24 h before infecting the animals. Following this, we assessed the intensity of bacterial phagocytosis by macrophages (Hamburger's index, Roitt's index) obtained from the abdominal cavity at various times using light microscopy. The highest intensity of bacterial phagocytosis by peritoneal macrophages was observed at the third, fourth, fifth and sixth hour after infecting the animals. MPs enhanced the phagocytic activity of macrophages at these intervals by 3.0; 2.59; 1.8; 1.43 times (Hamburger's index) and by 2.1; 2.47; 3.63; 5.88 times (Roitt's index). The differences with the control values were statistically significant in all cases ($p < 0.001$).

We also studied the influence of MPs on the intensity of elimination of injected live microorganisms *St. aureus* MT-1 from mouse abdominal cavity in the clearance test, that allows one to judge the efficiency of the macrophage barrier of the abdominal cavity. The results of the study of MPs effect as well as that of some other immunostimulators (like muramyldipeptide and blastolysin) on the phagocytic activity of peritoneal macrophages in BALB/c mice are shown in Fig. 10. The amount of colony-forming bacteria in the absence of immunomodulators was taken as 100% (control). The activation of macrophage phagocytic activity was assessed by a decrease in the number of colony-forming bacteria in the abdominal cavity of mice after the injection of the immunostimulator. MPs produce a pronounced stimulating effect on macrophage phagocytosis that exceeds the effect of all other immunomodulators used in the experiment. MPs induced the most intensive decrease in the amount of bacteria (Fig. 10).

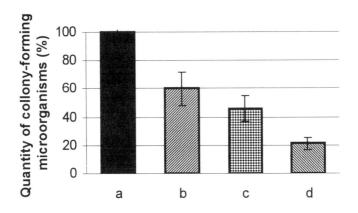

Fig. 10. Effect of immunostimulators on the phagocytic activity of peritoneal macrophages from BALB/c mice in the clearance test. (a) Control; (b) muramyldipeptides; (c) blastolysin; (d) MPs.

Stimulation of the functional activity of macrophages under the influence of MPs was also confirmed in the study on phagocytosis of mouse peritoneal macrophages in the nitroblue tetrazolium test (NBT test) (Rook *et al.*, 1985). Mice were infected with *S. typhimurium* 415. MPs were injected 24 h prior to infecting the animals. The results of the experiments showed that under the influence of MPs the bactericidal potency of peritoneal macrophages increased 1.55–1.73-fold as compared to the control. In this case a well pronounced protective effect developed, that resulted in a 100% survival of the animals, while in the control a 40–60% death rate was observed.

Therefore MPs isolated from the supernatant of porcine bone marrow cell culture contain compounds that augment the efficacy of immune reactions (enhancement of antibody formation, heightening of bactericidal effect etc.). This effect was displayed more intensively in those test animals which had a decreased level of the immune response. These data served as a starting point to a more detailed study of the ability of MPs to act as immunocorrectors, that is to correct immune impairments arising as a result of various pathological processes in the organism. If MPs are endogenous bioregulatory substances, we suggest that their action is directed towards normalization of the functioning of the impaired links of immunity.

## 2.2.2. Myelopeptides and Immunocorrection

It is known that the basis for the immune defense of an organism is a balanced functioning of multiple specialized immunocompetent cells that are concentrated in the central and peripheral organs of immunity. The immune system can be compared to a mobile with multiple sections that are extremely changeable but as a whole they are in an equilibrium state. The necessary balance between various sections of immunity is provided by a complicated network of interacting immunoregulatory signals. We suppose that MPs participate in the realization of these signals. This is evidenced by many experimental data demonstrating a significant ability of MPs to immunocorrection.

The immunocorrecting activity of MPs was revealed in various experimental models of immunodeficiency both *in vitro* and *in vivo*.

As one of the models of humoral immune response disorder with depression of antibody producers, we used MRL/lpr mice homozygous in the lymphoproliferation gene lpr. The presence of the lymphoproliferation gene in these mice causes autoimmune disorders like systemic *lupus erythematosus*. Apart from spontaneous autoimmune diseases, these mice also suffer from disorders in interleukin 2 (IL-2) production, higher T suppressor activity, high level of polyclonal activation and circulating immune complexes. MRL/lpr mice are characterized by an extremely low antibody production in response to SRBC immunization. In their lymph nodes, we could detect a minor amount of AFC on the fifth day after a secondary immunization with SRBC. However, if MPs were injected on the fourth day after the reimmunization — that is at the peak of the immune response — a high level of antibody formation was observed that could be compared to that in normal mice, *e.g.* in (CBA×C57BL)F$_1$ (Fig. 11). The level of circulating immune complexes and B cell polyclonal activation also were normalized under the influence of MPs.

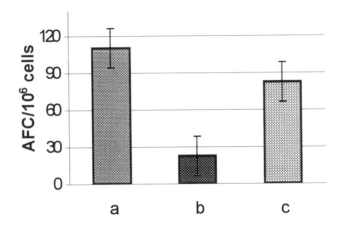

Fig 11. MPs correction of antibody response to SRBC in MRL/lpr mice. (a) Lymph node cells obtained from (CBA×C57BL)F$_1$ mice; (b) lymph node cells obtained from MRL/lpr mice; (c) lymph node cells obtained from MRL/lpr mice treated with MPs.

The congenital disorder of the immune status in MRL/lpr mice leads to an early death of such animals. Their life span is only five to seven months. A course of injections of MPs to these mice in the second, fourth and sixth months (100 μg/mouse thrice) increased their life span by 1.5 times (Mikhailova *et al.*, 1991).

It was noted that the bone marrow cells of MRL/lpr mice showed a significantly lower antibody-stimulating activity than that of normal mice.

When bone marrow syngeneic cells were added to the lymph node cell culture obtained from (CBA×C57BL)F$_1$ mice immunized with SRBC, the number of AFC/10$^6$ nucleated cells increased from 104±11 to 319±17. When bone marrow cells of MRL/lpr mice were added to the same immune lymph node cells obtained from (CBA×C57BL)F$_1$ mice, the AFC/10$^6$ number increased to only 205±17. It points to a deficiency of bone marrow cells of MRL/lpr mice in the production and/or functional activity of MPs. The exogenous administration of these immunoregulators can evidently compensate for the lack in MPs, thus ensuring a longer life span of the animals.

Aside from some congenital disorders, MPs can also correct immunodeficiency states caused by extreme environmental conditions. In particular, it was established that in mice under hypoxia (lifting the mice at 7,000 m by using the altitude chamber) the immune response to SRBC decreased by 30-50%. The administration of MPs to these animals on the fourth day of the secondary immune response normalized the processes of antibody production to SRBC. At the same time, a two- to threefold decrease in production of MPs by bone marrow cells in experimental animals was noted. The injection of MPs evidently leads to normalization of their level in the organism, that, in its turn, increases antibody formation up to the normal level.

A significant correction of the level of antibody formation under the influence of MPs was also observed in a model of immunodeficiency state in mice caused by stress and physical overload in the form of a dosed swimming load. This kind of stress caused suppression of the humoral immune response expressed as a 30-50% decrease in the AFC number in peripheral lymph organs of experimental animals. The main damaging factor under this stress is physical load. The swimming seances were conducted once a day under constant water temperature (30°C). The duration of the seance was 60 min.

Table 14 represents the data on the influence of MPs on antibody production in the lymph nodes of (CBA×C57BL)F$_1$ mice under swimming stress. Group 1 consisted of animals immunized with SRBC (control). Group 2 was composed of animals that had been subjected to stress: they received two swimming seances before the immunization (stressed control), in Group 3 the stressed animals received MPs two hours before the first swimming seance, and in Group 4 the stressed mice were injected with MPs on the fourth day after immunization with SRBC.

Table 14. Effect of MPs on the AFC number in the lymph node cell suspension obtained from stressed mice.

| Group No | Number of animals | $AFC/10^6$ nucleated cells | Coefficient of correction of antibody production | |
|---|---|---|---|---|
| | | | vs control | vs stressed control |
| 1 | 92 | 72.6±2.3 | | |
| 2 | 95 | 52.7±2.3 | 0.73 (p<0.001) | |
| 3 | 77 | 93.1±4.2 | 1.28 (p<0.001) | 1.77 (p<0.001) |
| 4 | 39 | 74.2±3.9 | 1.02 (p>0.05) | 1.41 (p<0.001) |

One can see that the administration of MPs prior to the stress prevented stress-induced decrease of antibody production (Group 3), since the AFC number in lymph node cells of these animals was higher than that of the control group (p<0.001). This fact points out to a prophylactic effect of MPs under stress-induced immunodeficiency.

When MPs were administered to the stressed animals at the peak of immune response to SRBC (Group 4), a positive effect of these regulators was also observed, in that the level of antibody production in the lymph nodes of these animals reached the control values and differed significantly from that in the group of stressed animals that had not received MPs (p<0.001) (Group 2).

There are some data concerning the influence of MPs on antibody formation in B cell hybridomas. A twofold increase in antibody synthesis was found after the addition of MPs to hybridoma 2B2 cells having a low level of antibody production to SRBC. At the same time, the addition of these peptides to hybridoma 2C12 cells that have a higher level of antibody production to this antigen did not change the level of antibody synthesis (Mikhailova *et al.*, 1987).

The immunocorrecting action of MPs is apparent not only in the system of antibody formation. It was demonstrated that MPs are capable of correcting the cellular immune response in a mixed culture of lymphoid cells. That is, MPs enhanced the formation of cytotoxic T lymphocytes in "a weakly immunogenic mixed culture" (with responder/stimulator cells ratio 20:1) of allogeneic mouse spleen cells. The addition of MPs to such a culture resulted in an increase in the cytolytic activity by 30-50%. At the same time, the addition of MPs to the normal mixed allogeneic cell culture (with responder/stimulator ratio 2:1) did not cause an increase in cytolytic activity (Derevyanchenko *et al.*, 1986).

The above mentioned examples illustrating the correcting action of MPs upon immune disorders point to the presence of immunoregulatory substances in the fraction consisting of a native mixture of nonidentified MPs. The action of these substances is directed towards the normalization of immune status disorders in the organism. Immunocorrection under the influence of MPs was observed both *in vitro*, and *in vivo*. When there were no immune disorders, the effect of MPs was less pronounced or totally absent; which is characteristic for most endogenous immunoregulators.

Obviously, MPs play a definite role in the preservation of normal immune homeostasis.

## 2.2.3. Differentiating Activity of MPs

The bone marrow is a hemopoietic organ that ensures the formation of all types of blood cells. It determines an extreme heterogeneity of bone marrow cell composition and its polyfunctionality. The development of different cell forms in the bone marrow requires the participation of some mediators that play an inductive and regulatory role in hemo- and immunopoietic processes, including cell differentiation. That is why we applied models of cell differentiation processes to test biological activities of the bone marrow mediators.

It is known that the entry of a cell into the differentiation process is accompanied by considerable changes in cell metabolism; the most typical one is a decrease in the synthesis of chromosomal DNA (that is, a decrease in cell proliferation) and the simultaneous increase in the synthesis of total proteins (that is, cell maturation). These parameters were used to assess the influence of MPs on the differentiation of mouse bone marrow cells *in vitro*.

The incubation of mouse bone marrow cells with MPs for one to two days resulted in a significant change in the synthesis of chromosomal DNA and total protein (without histones) which was measured by means of $^3$H-thymidine and $^{14}$C-glycine incorporation, respectively. In cultures incubated without MPs, DNA synthesis prevailed over protein synthesis, but in cultures incubated with MPs, we observed the predominance of protein synthesis over DNA synthesis. These data point to the fact that under the influence of MPs, the number of actively proliferating blasts in the population of bone marrow cells decreases and more mature cells appear (Strelkov *et al.*, 1989).

Especially, this regularity manifested itself when we investigated bone marrow and peripheral blood cells obtained from patients with hemoblastoses (Table 15).

Table 15. Effect of MPs *in vitro* on the ratio of DNA and total protein synthesis in bone marrow and peripheral blood cells obtained from patients with hemoblastoses.

| Cell donors | $^3H/^{14}C$ ratio | | | |
|---|---|---|---|---|
| | Peripheral blood cells | | Bone marrow cells | |
| | Control | MPs | Control | MPs |
| Acute myeloblastic leukemia | 0.56 | 0.40 | 15.3 | 8.8 |
| Acute lymphoblastic leukemia | 0.17 | 0.14 | 5.8 | 4.4 |
| Lymphosarcoma | 0.16 | 0.16 | 2.3 | 2.3 |
| Healthy | 0.15 | 0.15 | – | – |

The most marked changes in the ratio of DNA to protein synthesis under the influence of MPs were observed when bone marrow and peripheral blood cells from patients with acute myeloblastic leukemia were used. However, less dramatic changes were registered when cells from patients with acute lymphoblastic leukemia were used. MPs did not influence the ratio of DNA to protein synthesis when bone marrow and peripheral blood cells from lymphosarcoma patients, or peripheral blood cells from healthy donors, were used.

Hence MPs influence the metabolism and, consequently, cell differentiation in myeloblastic leukemia, that is, in hemoblastosis of bone marrow origin, more strongly than in other states.

The differentiating effect of MPs on leukemia cells was well reproduced in the human myelomonoblastic leukemia HL-60 cell line. The effect was assessed by measuring a decrease in DNA synthesis and a change in the expression of surface differentiation antigens revealed by means of monoclonal antibodies and morphological assessment of the amount of mature cells (monocytes/macrophages). MPs were incubated with HL-60 cells for 2 days. In parallel tests we used the known T cell-differentiating factor (TDF) reported by Leung and Chiao (Leung and Chiao, 1985) or recombinant interleukin 1 (rIL-1) instead of MPs.

Table 16. Effect of MPs on differentiation of the HL-60 cell line.

| Added agent | $[^3H]$ DNA (% vs control) | Differentiation antigens (marker-bearing cells, %) | | Morphology (cells, %) | | | |
|---|---|---|---|---|---|---|---|
| | | BMA 0200 | OKM 5 | Blasts | Promonocytes | Monocytes | Macrophages |
| Control | 100 | 96 | 2 | 90 | 10 | 0 | 0 |
| MPs | 58 | 75 | 47 | 50 | 20 | 25 | 5 |
| TDF* | 37 | 45 | 67 | 20 | 20 | 50 | 10 |
| rIL-1** | 95 | 89 | 17 | 75 | 25 | 0 | 0 |

*TDF, T cell-differentiation factor obtained from PHA-stimulated T cells (Leung and Chiao, 1985); **rIL-1, recombinant interleukin 1.

The data presented in Table 16 show that MPs cause a well pronounced differentiation of mono- and myeloblasts in the monocyte pathway. Thus incubation of HL-60 cells with MPs resulted in a twofold decrease in the synthesis of chromosomal DNA, as well as phenotypic changes characteristic for cell differentiation in the monocyte pathway and the appearance of a considerable amount of mature monocytes and macrophages (up to 30%) at the end.

The differentiating effect of MPs is similar to that of T cell-differentiating factor. In contrast, IL-1 has no differentiating effect in this model. The HL-60 cell line therefore proved to be a very suitable test system to estimate the differentiating activity of MPs.

To study the influence of MPs on cell proliferation and differentiation in the bone marrow in detail, we assessed the action of MPs on cell proliferation both in total and in fractionated bone marrow from (CBA×C57BL)$F_1$ mice. We did not succeed in revealing any changes in [$^3$H]-thymidine incorporation into DNA after 4 to 75 h joint cultivation with MPs in a total bone marrow cell population. At the same time, MPs exerted considerable influence on the proliferation of specific cell subpopulations. The bone marrow cells were fractionated by successive separation on plastic (removal of adherent cells), then on dishes loaded with an IgG$^+$ fraction from rabbit serum against mouse Ig (removal of IgG$^+$ cells). This cell suspension enriched in T lymphocyte precursors was then successively treated with antisera to remove Thy-1,2$^+$ and Sc-1$^+$ cells.

The incubation of specific bone marrow cell fractions with MPs for 16-18 h showed that MPs suppress the proliferation of adherent cells and cells of Thy-1,2$^-$ and Sc-1$^-$ phenotypes that belong to different hemopoiesis pathways. MPs stimulate the proliferation of Ig$^+$ cells in the bone marrow, that is, B lymphocytes as well as T lymphocyte precursors on two maturation stages with Thy-1,2$^-$, Sc-1$^+$ and Thy-1,2$^+$ and Sc-1$^+$, respectively. However, MPs had no influence on the proliferation of mature T lymphocytes with Thy-1$^+$ and Sc-1$^-$ phenotypes. The maximal proliferating effect of MPs was revealed on late T lymphocyte precursors (Thy-1,2$^+$ and Sc-1$^+$ phenotypes) that are active "helpers" in hemopoiesis (Petrov *et al.*, 1989).

Apart from the influence on the proliferation of specific bone marrow cell subpopulations, we established an influence of MPs on the early stages of T cell differentiation in the bone marrow. Like thymus hormones, MPs change the number of Thy-1,2$^+$ and Sc-1$^+$ cells and the amount of Thy-1,2 and Sc-1 antigens on the surface of bone marrow cells. Hence MPs participate in the development of bone marrow T lymphocyte precursors. They apparently determine the expansion and maturation of bone marrow progenitor T lymphocytes and together with thymus hormones (that reach the bone marrow from the blood) they prepare these cells for migration to the thymus, that is the organ responsible for the development of the T system of immunity.

An extremely important aspect of MPs action on bone marrow cells is their ability to induce the passage of early T lymphocyte precursors on the Thy-1,2$^+$ and Sc-1$^+$ cell stages which are the most active helpers in hemopoiesis. This effect of MPs apparently forms the basis for the mechanism of their regulatory influence upon hemopoiesis that is especially pronounced in hemopoietic disorders.

One of the examples illustrating the correcting effect of MPs on the impaired lymphopoietic processes is their participation in postirradiation recovery of the thymus. It is known that the regeneration of the thymus after irradiation is limited by proliferation, differentiation and migration of T lymphocyte precursors. The regeneration of the thymus has two main stages: the first one is at the level of intrathymus T cell precursors, the second one is at the expense of T cell precursors migrating from the bone marrow.

A single intraperitoneal injection of MPs to mice at the optimal dose ($10^{-8}$ g/mouse) 1 day after irradiation of the animal at a dose of 9.5 Gy considerably changes the course of thymus regeneration. MPs augment both the first and the second stages of regeneration that is measured by several parameters: cellularity of the organ, thymocyte proliferation *in vitro* and *in vivo,* changes in the expression of Thy-1,2 and Sc-1 antigens and receptors for peanut lectin (Kuznetsova, 1982). The participation of MPs in the postradiation recovery of the T cellular immunity along with their regulatory influence upon hemopoiesis apparently accounts for the prolonged life span of irradiated animals. The survival curves for mice after their lethal irradiation are shown in Fig. 12.

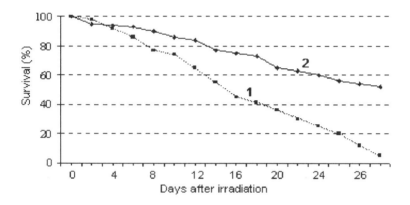

Fig. 12. Influence of MPs on the survival of irradiated mice. (1) - control; (2) - treatment with MPs.

The regulatory role of MPs in hemopoiesis can also be illustrated by our experiments on $W^v/W^v$ mice having spontaneous genetically predetermined anemia. Starting from the fourth to the fifth month of their life, such mice show a progressive fall in hemoglobin level, that results in the death of the animals. If MPs were injected to the animals simultaneously with the first manifestation of hemoglobin fall, then after one to three injections, the mice showed a gradual increase in the hemoglobin level. We managed to normalize the hemoglobin level in most animals (Fig. 13). To maintain this effect, additional supporting injections of MPs with ten days interval were necessary.

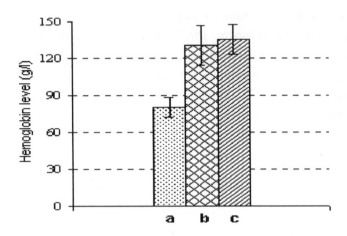

Fig. 13. Change in hemoglobin level in the peripheral blood of anemia $W^v/W^v$ mice treated with MPs. (a) Before the treatment with MPs; (b) after the treatment with MPs; (c) normal C57BL/6 mice.

The effect of MPs on proliferation and differentiation of hemopoietic precursor cells was also studied in $W^v/W^v$ mice (Petrov *et al.*, 1990). Some of them were injected with a single dose of MPs (300 µg/mice); the others were injected twice at doses of 300 and 100 µg/mice, with three days interval.

The bone marrow cells obtained from every $W^v/W^v$ mouse were injected intravenously into ten C57BL/6 mice irradiated at a dose of 7.3 Gy. Seven to eight days after cell transplantation, the spleens of the recipients were removed and fixed. After standard histological procedures, we obtained serial sections in which the number and type of the formed microcolonies were determined. The results of the experiments are shown in Fig. 14.

An injection of MPs induced a 2.5- to threefold increase in the number of granulocytic colonies and a nine to tenfold increase in the number of mixed erythroid-granulocytic colonies. The number of erythroid colonies either did not change or increased tenfold. Thus MPs can affect the formation of bone marrow hemopoietic precursor cells by correcting genetically predetermined defects that cause the development of macrocytic anemia.

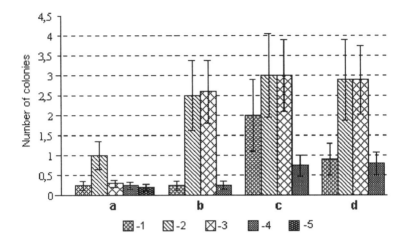

Fig. 14. Effect of MPs on the number and type of hemopoietic microcolonies formed by $W^v/W^v$ mouse bone marrow cells in the spleen of irradiated recipient. (1) Erythroid; (2) granulocyte; (3) mixed; (4) undifferentiated; (5) megacaryocute; (a) without MPs; (b) MPs (300 µg/mouse); (c) two injections of MPs (300 and 100 ·µg/mouse); (d) bone marrow cells from normal C57BL/6 mice.

The data obtained demonstrate various possibilities for accomplishment of the differentiating activity of MPs. These bone marrow mediators obviously participate in the processes of cell differentiation that occur in the bone marrow. This suggestion is confirmed by data on metabolic shifts that occur in the bone marrow cells from normal mice incubated with MPs, as well as the effect of MPs on the differentiation of early T lymphocyte precursors in the mouse bone marrow. However, the most pronounced effect of MPs on the differentiation processes is observed under hemopoiesis defects (hemoblastoses, macrocytic anemia). In such cases, the action of MPs is directed towards correction of these disorders, thus demonstrating the regulatory role of MPs in hemopoiesis.

The analysis of the data obtained in the study on the effect of MPs on cell differentiation suggests that the mixture of MPs contains several substances responsible for various differentiation effects. However the possibility of a single peptide from an MP mixture influencing the differentiation of various cell types in the bone marrow cannot be ruled out, and the form of its manifestation depends on the expression of the receptors to the given peptide on a definite target cell. To answer this question, it is necessary to isolate individual MPs.

## 2.2.4. Neurotropic Effects of Myelopeptides

In the early 1980s we conducted some experiments in collaboration with some physiologists and obtained results that were totally unexpected by us. In various experimental models it was shown that MPs have a neurotropic activity — they modify the pain sensitivity of the animals and influence their behavioral reactions (Petrov et al., 1982, 1987b). Knowledge on the role of immune system mediators in neuroimmune interactions was rather limited at that time. So the finding that immunoregulatory factors influence the nervous system evoked great interest.

The neurotropic effect of MPs was first revealed in our experiments on registration of evoked potentials in the brain cortex of large hemispheres in cats under electric stimulation of the tooth pulp (nociceptive stimulation) or of the lower lip (tactile stimulation). Intravenous injection of MPs (100 mg/kg) significantly decreased the amplitude of the evoked potentials under nociceptive stimulation and increased the amplitude under tactile irritation (Fig. 15). The effect was more pronounced in the frontal region of the cortex, that is responsible for emotional and behavioral reactions as compared to the second somato-sensory region. The effect lasted for 90 min and reached its maximum around 50–60 min after MPs administration, when the amplitude of the evoked potentials in the frontal region of the cortex was

55% of the initial level under nociceptive stimulation and 178% of that under tactile stimulation (Fig. 15). In the control animals (cats were injected with physiologic saline) the deviations in the amplitude of the evoked potentials as compared to the initial level did not exceed 5% during the whole experimental period. Naloxone injection to test animals 10 min before MPs abolished the changes in the amplitude of the evoked potentials.

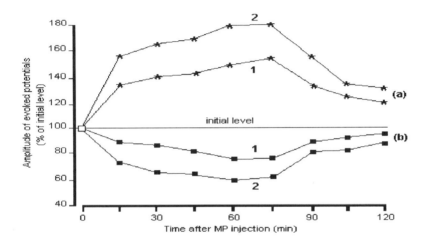

Fig. 15. Effect of MPs on evoked potentials in the cerebral cortex of cats under (a) tactile and (b) nociceptive stimulation. (1) the somatosensory region of the cortex; (2) the frontal region of the cortex.

The suppression of bioelectrical reactions in the brain cortex in response to nociceptive irritation and their augmentation in response to tactile stimulation together with the fact that this phenomenon is predominantly displayed in the frontal cortex region, is characteristic of narcotic analgesics like morphine and endogenous opiates endorphines.

The analgesic effect of MPs was observed in healthy volunteers who were injected subcutaneously with MPs at a dose of 0.075 mg/kg. The experiments were carried out on seven volunteers; two of them were injected with physiologic saline instead of MPs (control). The analgesic effect of

MPs was assessed by deviations in the amplitude of evoked potentials in response to electrocutaneous irritation of the forearm. The comparison of test and control data showed that an injection of MPs raised the sensitivity threshold by 47.5% on average, pain threshold by 37% and pain tolerance by 31.5%. The analgesic effect of MPs reached its maximum on 40 min after the injection of MPs. One hour after the administration of MPs, the evoked potential amplitude was partly restored, though it still deviated from the initial one at the end of the entire observation period (1.5 h).

The analgesic effect of MPs was also revealed in a behavioral model. For such a model we used the routine tail flick reaction of rats in response to nociceptive stimulation caused by a focused light beam. The latent period of nociceptive reaction was measured before and during 10 min intervals after the intraperitoneal injection of MPs at doses of 7.5 and 20 mg/kg. Control animals were injected with physiologic saline. The maximal peak of the analgesic effect of MPs was registered 40 min after the injection of MPs at a dose of 20 mg/kg (Fig. 16). Henceforth the activity of MPs slightly diminished but still it was significantly higher as compared to the control level at the end of the 1.5 h period. Physiologic saline injection, or the injection of MPs at a dose of 7.5 mg/kg, did not cause such an effect (Fig. 16).

Interesting results were obtained with a model of pain syndrome evoked in rats by a generator of pathologically enhanced excitation in nociceptive receptors in the lumbar region of the spinal cord (penicillin application). The pain syndrome was evaluated by a 3-mark system of six indices: (1) frequency of pain attacks; (2) duration of each attack; (3) intervals between the attacks; (4) response to provocation; (5) vocalization; and (6) mobile response.

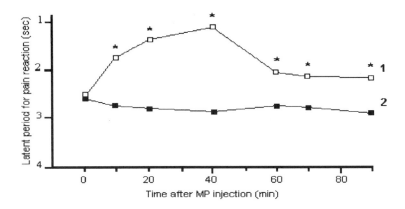

Fig. 16. Latent period enhancement of tail flick reaction in rats under the influence of MPs. (1) Treatment with MPs; (2) control; *p<0.05.

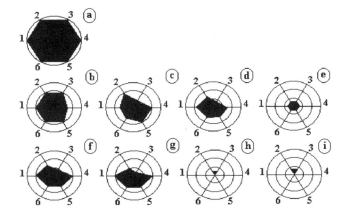

Fig. 17. Analgesic effect of MPs under pathologically enhanced excitation in the lumbar region of rat spinal cord. Pain evaluation indices: (1) frequency of pain attacks; (2) duration of each attack; (3) intervals between the attacks; (4) response to provocation; (5) vocalization; (6) mobile response; (a) maximal pain syndrome; (b–e) MPs doses 20, 30, 40, 50 mg/kg, respectively; (f) MPs (50 μg/kg) + naloxone (0.5 mg/kg); (g) analgin; (h) morphine; (i) promedole.

MPs were injected intravenously at doses of 20, 30, 40, 50 mg/kg at the peak of pain syndrome, that is at the moment when all its indices were at a maximum (Fig. 17a). The analgesic effect of MPs was dose-dependent (Fig. 17b-e), and at a dose of 50 mg/kg (Fig. 17e) it reached the effect of morphine (Fig. 17h) or/and promedole (Fig. 17i). The effect of MPs was even more pronounced than the effect of analgin (Fig. 17g). Naloxone injection to rats at a dose of 0.5 mg/kg just after the administration of MPs at a dose of 50 mg/kg partly abolished the analgesic effect of MPs (Fig. 17f).

It should be stressed that unlike morphine and promedole, MPs at the doses tested did not evoke either myorelaxing or narcotic effect, nor did they influence behavioral reactions not connected with the pain syndrome.

The data presented illustrates the naloxone-dependent antinociceptive opiate-like effect of MPs. In this regard it was expedient to study the ability of MPs to bind to opiate receptors. We analyzed the binding of MPs using the method of competitive replacement of [$^3$H]-morphine or [$^3$H]-met-enkephalin by MPs.

It was established that MPs have an ability to bind specifically with opiate receptors on rat brain cells as well as on human lymphocytes. It should be noted that MPs react better with opiate δ-receptors, than with opiate μ-receptors (Petrov et al., 1983-b).

These experimental results demonstrate the ability of MPs for specific binding with opiate receptors both on nerve cells and lymphocytes. Shortly thereafter it was shown that thymus peptides also bind with opiate receptors on rat brain cells (Zozulya et al., 1985).

Combining the evidence from the fields of psychology, neurobiology, and immunology has demonstrated that the immune system is not regulated exclusively in an autonomous fashion, but is influenced by factors directed by the central nervous system (Ader et al., 1991). On the other hand, there is a strong body of evidence suggesting that immune system signaling and activation are communicated to the central nervous system. This communication is thought to be realized through soluble factors released by cells of the immune system (Madden and Felten, 1995). Immunocompetent cells can synthesize and respond to most (if not to all) neuropeptides (Apte *et al.*, 1990, Wybran *et al.*, 1987; Gilmore and Weiner, 1988). Nerve cells produce many cytokines and respond to them (Bandtlow *et al.*, 1990; Minami *et al.*, 1990; Billiau *et al.*, 1989). The structural similarity of the receptors has been shown, *e.g.* for ACTH, endorphins, IL-1 and IL-2 (Blalock, 1989).

In light of recently published data, the phenomenon of neurotropic activity of MPs became understandable. Being immunoregulatory bone marrow mediators, MPs apparently participate in the neuroimmune interactions to provide certain signals in the complex network of intersystem links. This was clearly shown in our experiments on assaying the influence of MPs on pain sensitivity and antibody production in the same animals (Zakharova *et al.*, 1990).

The mice were intraperitoneally injected with various doses of MPs. The control animals obtained physiologic saline. At various time intervals after the injection of MPs (from 15 min up to three hours), the threshold of pain sensitivity in mice was determined in a hot plate test (Ankier, 1974). Every animal was tested once. 24 h after the administration of MPs and pain sensitivity assay, the mice were immunized with SRBC and on the fifth day the AFC number was determined in the spleen. The results of the influence of MPs on the immune response and pain sensitivity in these mice are presented in Fig. 18.

One can see that the influence of MPs on pain sensitivity in animals is dose-dependent. Low doses of MPs (from $1 \times 10^{-10}$ to $1 \times 10^{-6}$ g/mouse) decrease the latent period of the nociceptive reaction by 20–40%. The hyperalgesic effect of these doses was displayed 15 min after the injection of MPs  and lasted for another two hours (Fig. 18a). High doses of MPs ($7.5 \times 10^{-4} - 3 \times 10^{-3}$ g/mouse) caused an opposite, analgesic effect also lasting for two hours after the injection of MPs (Fig. 18b). Intermediate doses of MPs ($1 \times 10^{-5} - 2.5 \times 10^{-4}$ g/mouse) did not cause any change in pain sensitivity.

Assay of AFC number in the spleens of the same mice showed close correlation between the change in pain sensitivity and the intensity of the immune response. The lowering of the threshold of pain sensitivity under the influence of MPs was accompanied by a three- to ninefold increase in antibody production (Fig. 18a). The doses of MPs causing hypoalgesia did not provoke a statistically significant stimulation of the immune response (Fig. 18b).

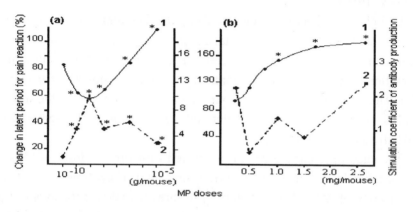

Fig. 18. Effect of MPs on pain perception (1) and antibody production (2) in mice. (a) High doses of MPs; (b) small doses of MPs; *p<0.05; **p<0.01; ***p<0.001.

The results obtained in this experiment indicate that there is a correlation between the immunoregulatory and neurotropic effects of MPs. However the use of the mixture of MPs does not allow us to assert with confidence that the same substances are responsible for these two different effects. It is not ruled out that our mixture of MPs consists of several components with different biological activities. The display of either hypo- or hyperalgesic effect depending on the concentration of MPs is more likely explained by the presence of substances having an opposite influence upon pain sensitivity. Either of the effects is displayed only under optimal concentration of the corresponding molecules. These questions could only be answered when individual, active MPs could be isolated and identified.

## 3. Myelopeptides and Human Immune Reactions

The experimental data presented in the preceding sections show that MPs have a wide spectrum of biological activities. These mediators are produced by bone marrow cells of various animal species and man and apparently participate in the complex network of bioregulatory processes.

It was therefore interesting to ascertain the role of MPs in the human immune and hemopoietic systems and their possible use to correct immune and hemopoietic disorders.

In this chapter, we present some data on the effect of MPs on the human immune system and summarize the results on clinical application of the developed medicinal preparation, Myelopid, whose action is due to MPs.

### 3.1. Influence of Myelopeptides on the Functional Activity of Human Peripheral Blood Cells from Healthy Donors and Immunodeficient Patients

The data obtained from studies on the influence of MPs on immune reactions in experimental animals show that these peptides manifest their maximal effect under immunodeficiency conditions. Thus we studied the action of bone marrow peptides on immunocompetent cells isolated from peripheral blood of healthy donors and immunodeficient patients. The immunocorrecting effect of MPs on peripheral blood cells from patients with impaired immune status can be illustrated by the action of MPs *in vitro* on the ratio of regulatory T cell subpopulations (CD4/CD8, helper/suppressor) in the blood of septic patients (Skryabina *et al.*, 1987).

This pathology is characterized by a shift in CD4/CD8 ratio towards CD8 prevalence. The incubation of blood cells obtained from such patients with MPs for 14–18 h resulted in considerable normalization of this pathological shift. The effect was most pronounced in newborns with sepsis who displayed sharp shifts in CD4/CD8 cell ratio (Table 17). When tested with blood cells from healthy donors, MPs displayed practically no effect on the helper/suppressor ratio of T lymphocytes.

Table 17. Change in CD4/CD8 cell ratio in blood of septic patients
after incubation with MPs *in vitro*.

| Groups | Incubation conditions | Healthy donors | | | Septic patients | | |
|---|---|---|---|---|---|---|---|
| | | CD4 | CD8 | CD4/CD8 | CD4 | CD8 | CD4/CD8 |
| Adults | without MPs | 40.2±0.9 | 14.2±0.8 | 2.8 | 22.3±0.8 | 32.3±0.4 | 0.6 |
| | with MPs | 42.7±0.7 | 13.8±0.7 | 3.0 | 32.8±0.9 | 24.8±0.9 | 1.3 |
| New-borns | without MPs | 32.2±0.3 | 38.6±0.6 | 0.8 | 14.0±0.2 | 58.6±0.3 | 0.2 |
| | with MPs | 36.4±0.4 | 32.4±0.9 | 1.1 | 28.4±0.8 | 40.3±0.8 | 0.7 |

It is known that complex surgical operations (surgical traumas) often cause various secondary immunodeficiency states. They are typical of patients with heart diseases developed as a consequence of rheumatic conditions. The immunodeficiency state formed in the course of the disease is aggravated during surgery and early postoperative stages (one to fourteen days after the operation) thus causing numerous infectious complications. The secondary immunodeficiency state in these patients can manifest itself by a lower functional activity of immunocompetent cells isolated from their peripheral blood.

To assess the ability of bone marrow peptides to influence the functional activity of immunocompetent cells obtained from postoperative patients, the model of spontaneous and pokeweed mitogen (PWM)-induced Ig synthesis *in vitro* was used. The effect of MPs on IgM, IgG and IgA production was studied (Jahn *et al.*, 1988).

Mononuclear cells were isolated from peripheral blood of postoperative patients and incubated with MPs. After 24 h incubation, the Ig level in the cultures was determined (basal Ig synthesis). It was established that one or two days after complex surgery the ability of mononuclear cells to carry out basal synthesis of all three Ig classes decreases by 1.5- to twofold as compared to basal Ig synthesis by healthy donor cells. Later, on the seventh or eighth day after the operation, we noted considerable activation in basal IgM and IgG production. Their levels were comparable to those in healthy donor cell cultures. At the same time, the level of basal IgA synthesis remained rather low. Addition of MPs to both healthy donor cells and those obtained from patients at the early (the first/second day) or at the late (the seventh/eights day) stage after the surgery did not cause any change in basal Ig production.

Thus MPs had no effect on basal synthesis of all three Ig classes of interest either in healthy donors or in patients with secondary immunodeficiency states after the surgical trauma.

As to the influence of MPs on the PWM-induced Ig synthesis, the situation was quite different. We used various mitogen doses which allowed us to assess the stimulating activity of bone marrow peptides under conditions of higher and lower initial Ig synthesis activation. Figure 19 demonstrates IgG production by healthy donor cells as a function of PWM concentration.

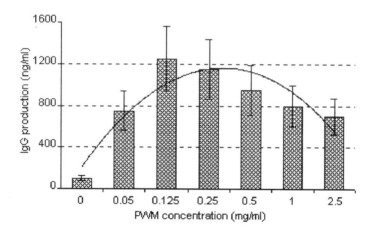

Fig. 19. IgG production by healthy donor peripheral blood cells *in vitro* as a function of PWM concentration.

It is evident that an increase in PWM concentration from 0.05 to 0.125 µg/ml resulted in a linear increase in IgG production in the cultures, the concentration of 0.05 µg/ml displaying an effect half as much as that of

0.125 µg/ml PWM. Further increase in mitogen dose above 0.125 µg/ml did not cause augmentation in IgG production. PWM concentration reduction below 0.05 µg/ml resulted in an unstable IgG synthesis in blood cells of various donors. Therefore doses of 0.05 and 0.125 µg/ml were selected for further investigation.

The study on the ability of selected mitogen concentrations to stimulate the synthesis of various Ig classes by patient blood cells obtained at various times after surgery showed that such stimulation was only found with cells isolated on the seventh or eighth day after the operation. Stimulation of cells obtained on the first or second day after surgery did not cause an increase in Ig production *in vitro*.

Table 18. Effect of MPs on the Ig synthesis in human blood cell culture under stimulation with the suboptimal PWM concentration (0.05 µg/ml).

| Cultivation conditions | Ig | Ig synthesis, ng/ml | | |
|---|---|---|---|---|
| | | Healthy donors | Postsurgery patients | |
| | | | $1^{st}/2^{nd}$ day after surgery | $7^{th}/8^{th}$ day after surgery |
| without MPs | IgM | 1222.9±180.8 | 236.4±42 | 252.1±87.4 |
| | IgG | 977.7±181.6 | 415.2±108.8 | 1215.8±256.8 |
| | IgA | 1050±175.4 | 394.8±150.6 | 2490.6±801.7 |
| with MPs | IgM | 1250.3±169.6 | 296.3±38.7 | 692.1±97.9 |
| | IgG | 1098.8±200.7 | 476.4±43.8 | 1563.7±279 |
| | IgA | 1746.6±335.7 | 440.4±76.2 | 5703.5±1132 |

The addition of MPs to cultures of healthy donor cells stimulated with a suboptimal mitogen dose (0.05 µg/ml) caused IgA synthesis enhancement, while in the cultures of cells from patients under the same mitogen doses, an increase in IgM and IgA production was found. But this was the case only when cells isolated on the seventh/eighth postsurgery day were used. Cells obtained on the first or second day after the operation were not sensitive either to MPs or to PWM alone (Table 18).

MPs did not cause any Ig synthesis stimulation in healthy donor blood cells when the mitogen was used at its optimal dose. In cells from immunodeficient patients, we found an increase in IgM and IgA production if blood cells were obtained on the seventh/eighth day after surgery. The effect of MPs in this case was less pronounced (Table 19).

Table 19. Effect of MPs on Ig synthesis in human blood cell culture under stimulation with the optimal PWM concentration (0.125 µg/ml).

| Cultivation conditions | Ig | Ig synthesis, ng/ml | | |
|---|---|---|---|---|
| | | Healthy donors | Postsurgery patients | |
| | | | $1^{st}/2^{nd}$ day after surgery | $7^{th}/8^{th}$ day after surgery |
| without MPs | IgM | 1921.6±244 | 301.2±161.5 | 1843.2±304.2 |
| | IgG | 1672.3±171 | 820.7±129.8 | 1246.8±263.2 |
| | IgA | 1497.7±175 | 394.8±150.2 | 1703.5±135 |
| with MPs | IgM | 2063.5±142 | 295.2±52.5 | 3188.7±517.1 |
| | IgG | 1693±275.7 | 804.3±127.1 | 1570±331.7 |
| | IgA | 1532±346.3 | 383±145.7 | 2490.6±101.7 |

The data obtained indicate that MPs stimulate Ig synthesis in activated cells only. They display no effect on basal Ig synthesis in cells of healthy donors or surgical patients. They do not affect Ig production in immunodeficient patient cells on the first/second day after the operation either, when the cells have lost their sensitivity to PWM. The maximal stimulating effect of MPs was found in cells isolated on the seventh/eighth day after surgery under stimulation by a suboptimal mitogen dose. The effect is of isotype-depending character as MPs cause an increase in IgM and IgA production, but not in that of IgG.

In the experiments presented above, MPs were added to the cultures simultaneously with the mitogen; their effect was assessed after seven days when significant Ig synthesis induction in the cultures could be measured. If MPs were added one day before the termination of the cultivation — that is six days after the onset of the cultivation with the mitogen — IgM and IgA synthesis stimulation was also observed (Fig. 20). Thus, the effect previously found with immune system cells of various animals was reproduced on human peripheral blood cells: MPs stimulate Ig production in the productive phase of the immune response.

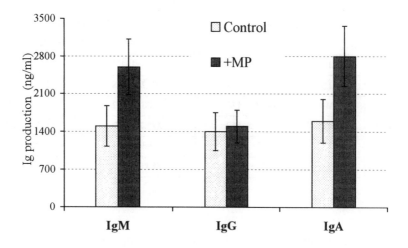

Fig. 20. Ig synthesis in cell cultures with MPs added six days after the onset of cell cultivation.

The data obtained in the studies on the effect of MPs on basal and induced Ig synthesis in cells isolated from the blood of healthy donors and patients with secondary immunodeficiency allowed us to arrive at the following conclusions:

1. The addition of MPs to both healthy donor cells and those obtained from immunodeficient patients on the first or second day and on the seventh or eighth day after an operation did not cause any change in basal Ig production.

2. After PWM induction of Ig synthesis, activated cells sensitive to MPs are generated that lead to an increase in IgM and IgA production in culture.

3. The level of the effect of MPs on Ig synthesis in PWM-stimulated cells is higher when suboptimal doses of the mitogen are used. It is also more pronounced with cells of immunodeficient patients. This higher effect of MPs on Ig synthesis in PWM-stimulated lymphocytes from patients with an impaired immune system was verified by a study on peripheral blood cells isolated from patients suffering from inherited hypogammaglobulinemia or selective IgA deficiency (Jahn *et al.*, 1988). The results are presented in Table 20.

Table 20. Effect of MPs on basal and induced Ig synthesis by peripheral blood cells obtained from patients with hypogamaglobulinemia and selective IgA deficiency.

| Hypogammaglobulinemia | | | | | |
|---|---|---|---|---|---|
| *Basal synthesis* | | | *PWM-induced synthesis* | | |
| *IgM* | *IgG* | *IgA* | *IgM* | *IgG* | *IgA* |
| – | 2.3* | 1 | – | 1.85 | 3.7 |
| *Selective IgA deficiency* | | | | | |
| *Basal synthesis* | | | *PWM-induced synthesis* | | |
| *IgM* | *IgG* | *IgA* | *IgM* | *IgG* | *IgA* |
| 1 | 1.3 | 1 | 1 | 0.9 | 2 |

* The ratio of Ig synthesis level in the cultures with MPs to those without it.

One can see that MPs stimulate the basal IgG synthesis and PWM-induced IgG and IgA synthesis in blood cells from patients with hypogammaglobulinemia as well as basal IgG and induced IgA production in cells from patients with selective IgA deficiency. At the same time, MPs cause no IgM synthesis stimulation in such cells. It is supposed that in blood of patients suffering from hypogammaglobulinemia and selective IgA deficiency, activated cells are present that are capable of IgG production and under the influence of MPs they begin to produce the Ig of this class. It is possible that the action of MPs on Ig synthesis is not isotype specific but is connected with immunoregulatory processes proceeding in the immune system under the participation of these peptides. This suggestion is substantiated by the data on MPs' inhibition of IgG synthesis in the presence of PWM in cells from patients with systemic *lupus erythematosus*. Such patients are characterized by a high level of basal IgG synthesis in peripheral blood cells. The cultivation of cells from these patients with PWM reduces the synthesis intensity to 58% of the control. Addition of MPs to the cultures causes further synthesis inhibition to 22% of the control — that is, the synthesis level in the cultures without mitogen and MPs. So, depending on the processes occurring in the immune system, MPs can produce either a stimulating or an inhibiting action on Ig synthesis in human peripheral blood cells, thus normalizing the processes of cellular interactions. Considered from this point of view, it becomes understandable why MPs affect the immune reaction in the course of its development, when activated cells are already present and the processes of cellular interactions reach the highest intensity. This is substantiated by study on the action of MPs upon mitogen-induced proliferation of cells isolated from peripheral blood of healthy donors. PWM, Con A, PHA are used as mitogens. MPs do not affect the level of $[^3H]$ thymidine incorporation when used alone. When used together with the mitogen, they cause an augmentation of proliferative response (Table 21).

Table 21. Effect of MPs on mitogen-induced [³H] thymidine incorporation
into mononuclears of peripheral blood from healthy donors.

| Mitogen | MPs | [³H] thymidine incorporation (cpm×10³) |
|---------|-----|----------------------------------------|
| Control | + | 0.86±0.11 |
|  | – | 0.98±0.24 |
| PWM | + | 20.74±2.29 |
|  | – | 13.16±1.34 |
| Con A | + | 22.00±2.52 |
|  | – | 15.38±0.92 |
| PHA | + | 34.10±2.60 |
|  | – | 24.88±2.30 |

If MPs really affect the interactions between various immunocompetent
cells in the course of immune reactions, then it should be accompanied by a
change in expression of surface cell markers (*e.g.*, CD25, HLA-DR, CD2
antigens). Figure 21 presents the results of a study on the action of MPs  on
the expression of these antigens in a culture of PWM-stimulated peripheral
blood cells.

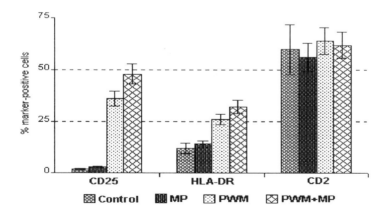

Fig. 21. Effect of MPs on the expression of surface markers on PWM-
stimulated human peripheral blood mononuclear cells.

The data shown in Fig. 21 demonstrate that 24 h cultivation of unstimulated peripheral blood mononuclears in the presence of MPs does not cause any change in the expression of surface markers. However, the addition of MPs to the PWM-stimulated cultures augments the expression of CD25 and HLA-DR antigen. CD2 antigen expression remains at the same level. CD4 and CD8 antigen expression does not change either (data not shown).

Thus, MPs do not activate silent cells, they join in regulatory processes in the course of the development of immune response. Under 24 h cultivation of mononuclear cells, MPs do not affect the expression of CD2 marker (T lymphocytes), CD4 (T helpers) or CD8 (T suppressors). At the same time, they increase the number of cells with HLA-DR and CD25 (IL-2R) antigens in the PWM-stimulated cultures. It is also possible that MPs consist of individual biologically active peptides whose action is directed at different links in the complex network of cellular interactions that arise as a response to the mitogen or antigen stimulus, thus providing for optimal development of immune reactions under given physiological conditions.

### 3.2. Influence of Myelopeptides on Bone Marrow and Peripheral Blood Cells Obtained from Healthy Donors and Patients with Some Hemopoietic Disorders

To understand the role of MPs in the functioning of the bone marrow, we addressed the question whether the production of these peptides in the bone marrow cells changes under some pathological states and whether the correction of impaired differentiation processes in the bone marrow by MPs is possible. One of the most widespread bone marrow pathologies are hemoblastoses. We studied the activity of MPs isolated from the bone marrow cells of healthy donors and patients suffering from acute myeloblastic leukemia, acute lymphoblastic leukemia, lymphosarcoma, chronic lymphocytic leukemia (myeloma disease) and agammaglobulinemia (Petrov et al., 1984).

First of all, we investigated the antibody stimulating activity of MPs produced by bone marrow cells isolated from such patients. As mentioned above, MPs have no species specificity (section 1.2). So we added MPs isolated from the cultural supernatants of bone marrow cells obtained from patients with various hematologic disorders to the lymph node cells obtained from mice at the peak of the secondary immune response to SRBC. After 20 h cultivation the stimulation coefficients were determined and compared to those in control cultures in which the MPs from healthy donor bone marrow cells were used. The results shown in Figure 22 demonstrate significant differences in the activity of MPs under study.

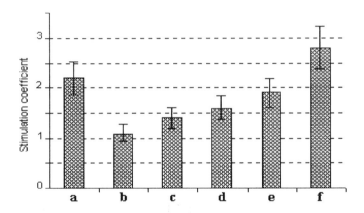

Fig. 22. Biological activity of MPs obtained from human bone marrow cells in normal and various pathological states. (a) Healthy donors; (b) agammaglobulinemia; (c) myeloblastic leukemia; (d) lymphoblastic leukemia; (e) lymphosarcoma; (f) chronic lymphocytic leukemia.

So, MPs isolated from healthy donors (Fig. 22a) can increase the number of antibody producers by $2.2 \pm 0.06$ fold, while the MPs obtained in the same manner from bone marrow cells from patients with acute myeloblastic leukemia (Fig. 22c), display a small stimulating effect on the number of antibody producers in the cultures of mouse immune lymph node cells (stimulation coefficient is 1.36±0.06). MPs contained in the culture supernatants of bone marrow cells from patients with acute lymphoblastic leukemia (Fig. 22d) and lymphosarcoma (Fig. 22e) are more active when added to the cultures of mouse immune lymph node cells (1.66±0.1 and 1.96±0.07, respectively). We could find practically no activity of MPs obtained from patients with agammaglobulinemia (1.07±0.03) (Fig. 22b). On the other hand, MPs produced by bone marrow cells of patients with chronic lymphocytic leukemia (myeloma disease) (Fig. 22f) have a more pronounced action than the peptides from bone marrow cells of healthy donors.

The myelograms of patients with acute lymphoblastic leukemia demonstrate that in the bone marrow of these patients there are many blast cells relating to the category of lymphopoietic precursor cells. The more pronounced the clinical symptoms of acute lymphoblastic leukemia are, the more inhibited the differentiation of stem cells to other hemopoietic pathways is. It is apparently accompanied by a decrease in the number of bone marrow MPs' producers, since a reduction in their activity is detected along with an increase in the number of lymphoblast cells. It should be noted that all patients with acute lymphoblastic leukemia suffered from the null-form of the disease (null leukemia) — that is, the lymphoblasts had neither T nor B cell markers on their surface.

The myelograms of patients with lymphosarcoma practically do not differ from those of healthy donors. The point is, that at the debut of the disease (the bone marrow cells were isolated from all these patients just after they were hospitalized, and before the onset of treatment) the tumor process does not affect the bone marrow. The lesion of bone marrow occurs along with lymphosarcoma progression and tumor process generalization, and the first atypical tumor cells are revealed in the myelograms. We investigated the bone marrow of patients with lymphosarcoma at the first stage of the disease, long before tumor process generalization.

The activity of MPs in patients with lymphosarcoma practically does not differ from the biological activity of these peptides in healthy donors

(Fig. 22a, e). The bone marrow producers of MPs are apparently not affected by the pathological process and continue to synthesize MPs.

We also studied the action of MPs isolated from healthy donors on the bone marrow and peripheral blood cells obtained from patients with the pathology of interest in order to establish their possible correcting action on the hemopoietic disorders available.

We assayed the effect of MPs *in vitro* on the integral indices of cell metabolism — chromosomal DNA and total protein synthesis — in the peripheral blood and bone marrow cells of patients with hemoblastoses. DNA synthesis was assessed by [³H] thymidine incorporation and protein synthesis by [¹⁴C] glycine incorporation. Some changes in these indices, namely a reduction in DNA synthesis along with a simultaneous increase in protein synthesis, indicate the onset of cell differentiation. Table 22 demonstrates the results obtained with cells from healthy donors and from patients before the treatment, just upon their hospitalization.

Table 22. Effect of MPs *in vitro* on DNA and total protein synthesis in peripheral blood and bone marrow cells obtained from patients with various hemoblastoses.

| Cell type | Hemo-blastosis form | [³H] thymidine and [¹⁴C] glycine incorporation / 10⁶ cells | | | | | | [³H] / [¹⁴C] | |
| | | [³H] DNA | | | [¹⁴C] protein | | | | |
| | | Control, imp/min | + MPs | | Control, imp/min | + MPs | | Control | + MPs |
| | | | imp/min in | % | | imp/min in | % | | |
| Peripheral blood cells | AML | 2200 | 1800 | 81 | 3900 | 4500 | 115 | 0.56 | 0.40 |
| | ALL | 810 | 680 | 84 | 4600 | 5000 | 109 | 0.17 | 0.14 |
| | LS | 660 | 650 | 100 | 4200 | 4200 | 100 | 0.16 | 0.16 |
| | Donors | 550 | 540 | 100 | 3700 | 3700 | 100 | 0.15 | 0.15 |
| Bone marrow cells | AML | 21800 | 14600 | 67 | 1420 | 1420 | 116 | 15.3 | 8.8 |
| | ALL | 3700 | 3200 | 87 | 640 | 640 | 113 | 5.8 | 4.4 |
| | LS | 1600 | 1580 | 100 | 700 | 700 | 100 | 2.3 | 2.3 |

AML, acute myeloblastic leukemia; ALL, acute lymphoblastic leukemia; LS, lymphosarcoma.

These data show that the incubation of peripheral blood cells from patients with acute leukemia with MPs causes a reduction in DNA synthesis (20% less than in the control on the average) and simultaneous increase in total protein synthesis (15% on the average). At the same time, MPs have practically no effect on healthy donor cells or on cells from patients with lymphosarcoma.

The joint cultivation of bone marrow cells from the same patients with MPs resulted in an even more pronounced reduction in cell proliferative activity along with simultaneous increase in total protein synthesis. It shows even more pronounced effects of these peptides on the functional state of bone marrow cells obtained from patients with various forms of hemoblastoses. This action of MPs is more pronounced under acute myeloblastic leukemia than acute lymphoblastic leukemia.

The materials presented show that under some pathological states, *e.g.* hemoblastoses, the production of MPs and/or their functional activity in the bone marrow changes significantly. The defects in cell differentiation in the bone marrow result in a dramatic change of its cellular composition. It evidently has an effect on the production of MPs. The differentiating ability of MPs found upon addition of MPs to bone marrow and peripheral blood cells from patients with acute myeloblastic leukemia (Table 22), indicates that these peptides participate in hemopoietic processes in the bone marrow. The fact that MPs cause a reduction in DNA synthesis (proliferation) and an increase in total protein synthesis (differentiation) in myeloblast cells allows one to consider these peptides as potential antileukemia agents.

### 3.3. Myelopid: A Medicinal Preparation Developed on the Basis of Native Myelopeptide Mixture

Our experimental data on the biological activities of MPs demonstrate that they have pronounced correcting properties. They were shown to normalize the impaired links of immuno- and hemopoiesis and exert only a very slight influence on the normally functioning systems in the organism. The absence of interspecies restriction enables one to isolate them from bone marrow cells of various animals and to use them for immunocorrection in humans. MPs displayed their correcting effects *in vitro* in human blood and bone marrow cells in various immunodeficient states and in hematologic disorders. All these data prompted us to develop a new immunocorrecting medicinal preparation on the basis of MPs, which was named Myelopid. Below, we characterize this preparation and describe its pharmacological and toxicological properties and clinical efficiency.

Myelopid is a highly purified mixture of MPs with molecular mass of 500–3000 Da isolated from the supernatant of porcine bone marrow cell culture by a solid phase extraction method.

The pharmacological activity of Myelopid depends on the biological effects of the various MPs of which it is composed of. Like MPs, Myelopid stimulation of differentiation of immunocompetent cell precursors in the bone marrow of immunodeficient patients results in an increase in the number of B and T lymphocytes and phagocytic cells in peripheral blood. Along with its ability to affect the differentiation of immunocompetent cells in the bone marrow, Myelopid also stimulates the functional activity of T cells and monocyte/macrophage cells activated with antigens of various bacterial or viral nature (Stepanenko *et al.*, 1989; Abbakumov *et al.*, 1990; Petrov *et al.*, 1990a; Nesterova *et al.*, 1991). Restoration of the immune response after Myelopid application is not accompanied by any undesirable side effects.

Concerning the toxicological characteristics of Myelopid, it can be regarded as a practically nontoxic medicinal preparation. Its safety index (the ratio of the maximal sublethal dose to the maximal effective dose) exceeds 30.

Pathophysiological investigations of the organs isolated from experimental animals after prolonged Myelopid injections (30 days, daily) reveal neither pathological disturbances in their hemopoietic system, nor dystrophic, necrobiotic changes in parenchyma and stromal cells, nor inflammatory reactions in the liver, kidneys, adrenal glands, heart, bone marrow, hypophysis, thymus, spleen, mesentery and groin lymph nodes, testicles, thyroid gland, pancreas, gullet, stomach and large intestine. Our study of lymphoid organs revealed an increase in the mass and cellularity of the lymph nodes that drain the sites of Myelopid injection as well as a slightly reduced mass and cellularity of the thymus. This is apparently connected with the migration of immunocompetent cells from the central lymphoid organs to the peripheral ones.

Biochemical investigations of blood serum and 24 h urine of animals exposed to prolonged Myelopid administration revealed no significant deviations from the norm, and no changes were observed in the function of the liver and kidneys either.

The study that was conducted showed that the preparation has no carcinogenic or mutagenic activity, neither has it allergenic effects. Myelopid injections at doses equal to ten or 100 times therapeutic ones does not induce immediate or delayed hypersensitivity reactions.

Myelopid is compatible with antibiotics, hormone or chemotherapeutic preparations and other medicinal preparations that are used in the treatment of chronic infectious and inflammatory diseases as well as in oncology.

Data on the clinical efficiency of our preparation in patients with various pathological processes accompanied by immunodeficiency, as well as data on the effect of Myelopid on the immune status in these patients, indicate that it is an immunomodulator with normalizing action. Myelopid corrects the impaired links of immuno- and hemopoiesis without causing any effects on the normal processes in the body. The restoration of the immune status in these patients promotes their more rapid recovery.

Myelopid proved to be a highly efficient medicinal preparation to prevent postsurgery and posttrauma infectious complications (Abbakumov *et al.*, 1990; Bogomolova *et al.*, 1993; Stepanenko *et al.*, 1991), to treat chronic and sluggish respiratory tract diseases (Stepanenko *et al.*, 1989; Petrov *et al.*, 1990-a); chronic inflammatory and purulent inflammatory processes of various localization (including generalized ones) (Lisianyi *et al.*, 1991; Nesterova *et al.*, 1991), bacterial and viral infections (Osmanova and Gurarii, 1989; Melnik and Stepanenko, 1994), and allergic diseases (*e.g.*, corticosteroid-dependent bronchial asthma, atopic dermatitis) (Zoloedov *et al.*, 1995). It is also used in the treatment of patients with acute lymphoblastic and acute myeloblastic leukemia and non-Hodgkin's lymphoma (Zotova *et al.*, 1990).

Nowadays, Myelopid is used in clinics to prevent general and local infectious complications after surgery, severe mechanical traumas, thermal and chemical burns, X-ray and chemotherapy and hormone preparation treatment as well as a consequence of unfavorable ecological factors. As a result, the risk of infectious complications developing diminishes significantly. The preparation is recommended to be administered before and just after operations, after severe traumas, during and just after X-ray, chemo- or hormonotherapy. Myelopid application enables one to reduce the possibility of infectious complications developing by a factor of two to three.

Below, we illustrate the application of Myelopid in the treatment of some pathologies.

Myelopid was used in the treatment of patients with rheumatic heart disease, including those diagnosed to have bacterial endocarditis (Abbakumov *et al.*, 1990; Bogomolova *et al.*, 1993). They underwent surgical correction of heart diseases under artificial blood circulation and ice-chip cardioplegia. Since postoperative infectious complications usually develop four to twelve days after the surgery, Myelopid treatment was started the very next day after the operation. The preparation was injected on the first, third, fifth, sevenths and eighth days.

It should be noted that the course of postoperative complications in patients treated with Myelopid considerably differs from that in the control group of patients. They proceed with less pronounced intoxication and fever and only moderate leukocytosis. Purulent discharge from the operative wound was insignificant. Moreover, the clinical improvement in patients with operative wound pyesis, who were treated with Myelopid, was evident after 12–14 days, while in the control group it was observed after 17–21 days.

The analysis of the immune status parameters of these patients showed a statistically significant decrease in the relative numbers of T and B lymphocytes, CD4⁺ cells, reduction in IL-2 production by PHA-stimulated T lymphocytes as well as a decrease in IgM and IgA production by B lymphocytes in response to PWM stimulation. The level of T lymphocyte proliferative response to PHA and spontaneous [³H] thymidine incorporation also decreased.

Patients, who received traditional treatment, displayed a slight increase in their immune status parameters on the seventh/eighth day after the operation. In particular, an increased spontaneous [³H] thymidine incorporation into peripheral blood mononuclears and an enhanced IgM production after PWM stimulation were observed. These processes are apparently caused by restoration of the functional activity of the immune system at the early postoperative stage.

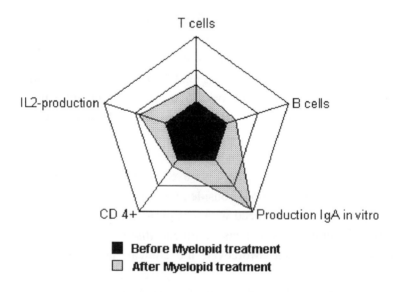

Fig. 23. Changes of the immune status parameters in patients with surgical correction of heart diseases after Myelopid treatment.

Immunocorrection with Myelopid, performed just after surgery, results in a more rapid restoration of the immune system parameters. Seven to eight days after the operation, the Myelopid-treated patients showed an increase in the relative number of T and B lymphocytes in their peripheral blood, an increase in the $CD4^+$ cell number, which resulted in an enhanced CD4/CD8 (helpers/suppressors) ratio as well as in a greater lymphocyte ability to increase IL-2 production following PHA stimulation. The spontaneous $[^3H]$ thymidine incorporation in these patients diminished. Figure 23 presents statistically significant differences ($p<0.05$) between the immune status of Myelopid-treated patients and of those in the control group.

Thus, Myelopid treatment of patients with secondary immunodeficiency at the early postoperative stage promoted restoration of some decreased immune status parameters. It positively affects the clinical course of the postoperative period.

Similar data were obtained in patients suffering from severe traumas. Infectious complications are known to develop rather often in these patients. Studies show that 40% patients with severe cerebral traumas develop pneumonia and purulent tracheobronchitis. Such a high level of infectious complications can be explained by T immunodeficiency that develops in patients after severe traumas as well as after extensive surgery. These are: T lymphopenia, impaired CD4/CD8 balance, reduced blasttransformation in response to PHA. Functional deficiency of the B system of immunity also occurs — the levels of serum Ig and immune complex decrease, autoimmune process develops, antiorgan antibodies and sensitized lymphocytes appear. A study of the immune status in patients with burns also demonstrates a drop in the functional activity of T and B lymphocytes and a CD4/CD8 disbalance. The level of immune disorders in this case depends on the severity of the trauma and on individual peculiarities of the organism. It correlates with posttraumatic inflammatory complications.

We studied the clinical efficiency of Myelopid in patients with various traumas (Stepanenko *et al.*, 1991). Addition of Myelopid to the complex treatment of patients suffering from traumas was shown to decrease the level

of purulent inflammatory complications by 1.5- to threefold. After four to
five injections, an improvement in the patient's clinical state, diminished
intoxication, higher efficiency of antibacterial therapy and an arrest of
purulent complications were found. Figure 24 presents the results of
Myelopid treatment of patients with fractures of the mandible and
accompanying diseases (alcoholism, chronic lung, liver and renal diseases,
diabetes) or without them.

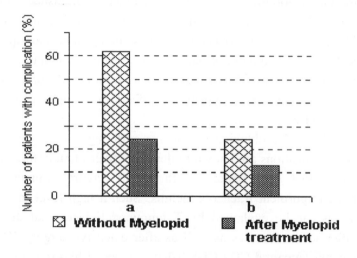

Fig. 24. Myelopid effect on the number of purulent inflammatory
complications in patients with mechanical traumas.

Myelopid administration to patients with traumas evoked an increase in
the number of T and B lymphocytes in their peripheral blood. We registered
a higher level of T helpers and a somewhat lower content of
T suppressors — that caused an increased helper/suppressor ratio (2.4 in
Myelopid-treated patients *versus* 1.19 in the control group).

Myelopid treatment of patients with traumatic osteomyelitis resulted in a
more rapid arrest of the inflammatory process not only in soft, but also in
bone tissues as compared to the group of traditionally treated patients. In

more than 50% patients, spontaneous withdrawal of sequesters or their resolution was found, thus allowing one to spare sequesterectomy. The hospitalization duration significantly decreased.

The data on the immune status parameters of patients with osteomyelitis show that these patients develop immunodeficiency states that remain at the same level after a course of traditional therapy, before their discharge from the hospital. On the other hand, patients additionally treated with Myelopid display positive alterations in the immune status, in particular, in CD4+/CD8+ ratio (Table 23).

Table 23. CD4+/CD8+ lymphocyte ratio in patients with osteomyelitis
after Myelopid treatment.

| | Before treatment | After treatment | |
|---|---|---|---|
| | | $7^{th}/8^{th}$ day | $14^{th}$ day |
| CD4 | 48.4±0.7 | 67.4±0.2 | 62.9±0.3 |
| CD8 | 36.9±3.5 | 27.8±0.8 | 29.4±2.6 |
| CD4/CD8 | 1.3 | 2.4 | 2.1 |

Myelopid incorporation into the complex treatment of traumatic mandible osteomyelitis causes an increase in the number of CD4[+] cells and some reduction in the number of CD8[+] cells. This is accompanied by other changes in the immune status, that are found after Myelopid therapy in patients at the early postoperative stage.

Less infectious diseases were registered in Myelopid-treated patients with thermal burns. Four or five injections of the preparation resulted in clinical improvement, reduced intoxication, increased efficiency of antibacterial therapy, and arrest of purulent complications. In the group of Myelopid-treated patients the number of purulent complications was decreased by a factor of 2.5 compared with the control group. Incorporation of Myelopid into the complex therapy of burns causes a higher survival of patients at the acute stage of the burn shock and septicotoxemia.

Thus the data outlined above show that Myelopid incorporation into the complex therapy of patients with various traumas results in correction of their immune status disorders, lower probability of general and local infectious processes, improvement of patients' clinical state and more rapid recovery.

Another area for Myelopid application is treatment of patients with chronic nonspecific respiratory tract diseases. The administration of the preparation to patients with chronic pneumonia in the acute periods of the disease caused improvement of their general state, stopped fever and sweating, lessened weakness, and also improved patients' sleep and temper. Steady remissions were registered in most patients treated with Myelopid. The duration of these remissions in 78% patients with chronic pneumonia was one year or more (Stepanenko *et al.*, 1989; Petrov *et al.*, 1990a, 1993).

Figure 25 presents the data on remission duration before and after Myelopid incorporation into the complex therapy of patients with chronic pneumonia.

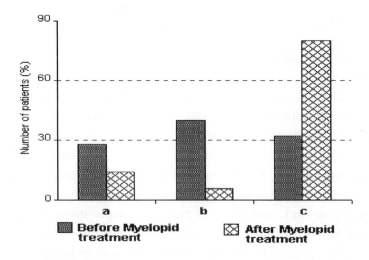

Fig. 25. Change in remission duration in patients with chronic pneumonia treated with Myelopid. Time of remission: (a) less than three months; (b) about six months; (c) over six months.

Before undergoing the Myelopid course, these patients with chronic pneumonia were characterized by a high level of recurrences. In spite of the intensive antibacterial therapy, they had exacerbation of respiratory tract process up to three to four times a year. This was often accompanied by exacerbation of maxillitis, tonsillitis, otitis, and ethmoiditis. The study of the immune status in these patients revealed significant individual variations in some parameters. As compared to healthy donors, 44% patients had considerably lower absolute numbers of T and B lymphocytes in peripheral blood; 23.8% patients displayed lower relative numbers of CD4+ lymphocytes; 22% had less CD8+ lymphocytes and in 17% patients a fall in the relative amount of CD11+ lymphocytes was registered.

The improvement in the clinical state of Myelopid-treated patients was accompanied by normalization of their immune status parameters. This effect was most pronounced ten days after the Myelopid treatment course. It can be illustrated with a 1.2-fold increase in the absolute number of B lymphocytes in the peripheral blood and a 15–20% increase in the number of CD4+ and CD11+ lymphocytes.

A multifactorial statistical analysis of Myelopid-induced changes in the immune status parameters shows that our preparation enhances the activity of T helpers and neutrophils and inhibits that of suppressor cells (Stepanenko et al., 1989).

We obtained analogous results when treating patients with chronic obstructive bronchitis who had up to five recurrences in one year. Traditional therapy proved to be inefficient in such cases. Myelopid incorporation into the complex treatment considerably improved the patients' clinical state. It was manifested by less intensive attacks of coughing and short breath. The clinical effect was steady and was accompanied by improved indices of outer breath in more than half of them. The study of the immune status parameters in this group of patients demonstrated an increase in the number of T lymphocytes in the peripheral blood and enhancement in T helper activity, as well as reduction in T suppressor activity.

Promising clinical results were obtained in a study of the effect of Myelopid on the course and outcome of appendicular peritonitis. Two ways of Myelopid administration were compared: subcutaneous and endolymphatic injections. The patients were divided into three groups. Group 1 was a control one (not treated with Myelopid). Group 2 received subcutaneous Myelopid injections, while group 3 was injected endolymphatically.

The immune status parameters were determined on the first day after surgery and at various times after Myelopid administration (on the fifth, ninth and fourteenth days).

On the first day after the operation the patients developed immunodeficiency; the immunograms in all three groups did not differ significantly. The distinctions in the immune indices in the groups were first noticed on the ninth/fourteenth day after the surgery.

On the ninth day all patients demonstrated reduced leukocytosis and lymphopenia was more pronounced in patients of group 2 and group 3. In group 3, the number of lymphocytes became normalized at this stage.

Furthermore, the patients in group 2 displayed a moderate increase in the relative number of B lymphocytes, a threefold increase in the IgM level and enhanced complement activity of blood serum simultaneously with a decrease in the number of circulating immune complexes.

At this stage, the patients in group 3 showed normal amounts of T lymphocytes and significantly increased absolute and relative amounts of B lymphocytes. IgG and IgM levels and complement activity of blood serum increased, and simultaneously the number of circulating immune complexes was reduced. The alterations in the immunograms of these patients were accompanied by less pronounced intoxication symptoms, complete restoration of gastroenteric tract function and by less pronounced symptoms of renal and hepatic failure.

On the 14th day, the patients in group 1 showed leukocytosis and lymphopenia; the amount of T and B lymphocytes and the level of Ig of all three classes remained less than normal.

In patients of group 2, the number of leukocytes and lymphocytes normolized. The number of B lymphocytes, IgM, circulating immune complexes and the complement activity of blood serum remained high.

The patients in group 3 also showed normal amounts of leukocytes and T lymphocytes. The absolute number of B lymphocytes normolized, but their relative amount and their functional activity were still high. The IgG level normalized, while the IgM content even increased. The complement activity of blood serum remained high.

The positive changes in the parameters of the immune status after complex treatment of peritonitis considerably improved the clinical course of the disease. The data illustrating the effect of Myelopid on the course and outcome of peritonitis are shown in Table 24.

Table 24. Myelopid treatment of patients with peritonitis.

| | Endolymphatic injection | Subcutaneous injection | Without immunocorrection |
|---|---|---|---|
| Mean duration of high temperature reaction (days) | 6.4 | 8.2 | 12.1 |
| Mean duration of intestinal poresis (days) | 2.3 | 3.8 | 4.2 |
| Postsurgery complications (%) | 10.0 | 26.6 | 47.3 |
| Hospitalization duration (days) | 16 | 23.5 | 30.1 |
| Lethality (%) | 5.0 | 20.0 | 27.0 |

Thus the immunodeficiency that develops in patients with peritonitis, is corrected by Myelopid treatment. Its incorporation in the treatment of such patients results in a more smooth postoperative period, faster normalization of body temperature and peristalsis restoration; it reduces lethality and decreases the number of postoperative complications. As a result, the hospitalization period is lessened. Endolymphatic administration of Myelopid results in a more rapid restoration of immune status parameters and a higher efficiency of immunomodulating therapy.

Myelopid application on patients with bacterial infections (dysentery, typhus and paratyphus, acute and subacute brucellosis, salmonellosis) shows, that the preparation decreases the symptoms of acute intoxication; the duration of hospitalization period is lessened. Myelopid prescription to patients with acute brucellosis enables one to achieve complete clinical remission at the hospitalization stage (Osmanova and Gurarii, 1989).

In patients suffering from tuberculosis, Myelopid immunotherapy in combination with antibacterial drugs diminished intoxication symptoms. Figure 26 demonstrates the dynamics in the closure of decay cavities — that is the main measure of therapy efficacy — in patients who were treated with Myelopid along with antibacterial therapy (group a) and those who were treated conventionally (group b).

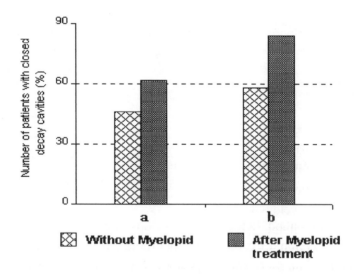

Fig. 26. Myelopid treatment of tuberculosis patients.

The comparative analysis of the immune status parameters in these groups of patients showed a statistically significant increase in the absolute and relative numbers of T and B lymphocytes in peripheral blood, enhancement in the proliferative response to PWM and phagocytic activity of neutrophils after Myelopid treatment.

We also have positive results of Myelopid treatment of patients with various diseases of viral nature: hepatitis, herpes and cytomegalia (Melnik and Stepanenko, 1994). Myelopid incorporation into the treatment of such patients results in a more rapid (by eight to fourteen days) normalization of the functional hepatic probes, cytochemical and immune indices

Patients suffering from corticosteroid-dependent bronchial asthma are treated with Myelopid on the background of routine therapy (broncholytics, hormones, cholinilytics, adrenomimetics, spasmolytics, expectorants and antihistamine medications). Myelopid treatment in this case results in normalization of expiration, finer rale and improvement of patients' state (Zoloedov, 1995).

Myelopid injection in patients suffering from alimentary allergy with atopic dermatitis having on the skin the signs of secondary infection showed therapeutical effect on the third/fourth day after the treatment course termination. It was manifested as less intensive inflammatory and infiltrative processes in the damaged skin regions, extinction of infection focuses in the skin and more intensive regeneration processes. Positive clinical results were registered in 67% patients. Prolonged catamnesis supervision indicates termination of pustule recurrences for six months.

Myelopid application in the treatment of hemoblastoses should also be discussed (Zotova *et al.*, 1990). It is known that pneumonia, angina and stomatitis often develop at the final stage of any hemoblastosis, but they can be the first clinical manifestations of this disease as well. Prescription of antitumor (cytostatic) preparations aggravates the impaired immunity in such patients and promotes to development of infectious complications and hemoblastosis recurrences. After chemotherapy, the patients are practically defenseless in front of various infecting agents; they often develop various infectious complications.

A single intraosteal or intramuscular Myelopid injection causes a hematological effect displayed as reduction in the relative blast number in the bone marrow (Fig. 27). The second injection does not augment this effect, the patient's state stabilizes—and this is the base for the onset of antitumor therapy. It is much easier for the patients to enter the remission stage. The infectious complications at the induction stage are much more rare.

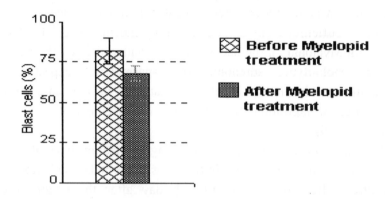

Fig. 27. Change in the number of blast cells in the bone marrow of patients with myeloblastic leukemia.

After the termination of the chemotherapy course, Myelopid was applied again. It was injected intramuscularly, five to ten injections every other day.

Morphological investigations of the bone marrow and peripheral blood from patients with acute leukemia, that were treated with Myelopid, show that a single intraosteal injection causes a short hematological effect — it causes a reduction in the absolute and relative number of blast cells and an increase in the number of cells, that according to morphological criteria, can be considered as lymphoid elements. This is accompanied by an increase in the number of B lymphocytes in the peripheral blood. The repeated Myelopid course just after patient's entering the remission stage resulted in an increase in the number of neutrophils.

Myelopid injections along with routine antibiotic therapy of patients with T and B cell non-Hodgkin's lymphomas at the third or fourth stage of the disease after intensive chemotherapy considerably promote a milder course of infectious complications (pneumonia, sepsis, peritonitis) and augment the absolute blood indices at shorter times as compared to patients who were not treated with Myelopid.

The formation of mature cell lymphoid elements in the peripheral blood of leukemia patients is typical of patients who were treated with Myelopid along with traditional cytostatic therapy, but not with cytostatic therapy alone. This fact indicates that apart from immunocorrection, Myelopid can induce blast differentiation to more mature cell forms — that is, it can act as a differentiation factor. In section 3.2, we presented some data on the ability of MPs to induce *in vitro* differentiation of bone marrow cells from patients with hemoblastoses. This differentiating activity of MPs was most pronounced in bone marrow cells obtained from patients with acute myeloblastic leukemia. It is interesting to note that bone marrow cells, that were obtained from leukemia patients after Myelopid course, reduce or even lose their sensitivity to MPs *in vitro*.

Myelopid injected to leukemia patients apparently influences the maximal number of blasts capable of differentiation. After Myelopid treatment, the number of bone marrow cells that can respond to the MPs' differentiating stimulus is very low. As a result, we registered a decrease in the MPs' sensitivity of bone marrow cells obtained from Myelopid-treated patients *in vitro* as compared to their sensitivity before Myelopid treatment.

It should be stressed that clinical application of Myelopid in more than 30,000 patients displayed no worsening in the state of any patient. This fact could be explained in terms of the endogenous nature of the peptides of which Myelopid is composed.

The result of a nine year clinical trial of Myelopid is summarized in Table 25.

Table 25.

| Myelopid treatment is efficient | No pronounced effect registered | Further studies are needed |
|---|---|---|
| 1. As prophylactic preparation to prevent infectious complications (pneumonia, bronchitis, wound purulence) <br><br>2. As medicinal preparation at the acute stages of chronic infectious and purulent processes in respiratory tract, gastrointestinal and urogenital systems. Chronic neuroinfections <br><br>4. In complex therapy of hemoblastoses <br><br>5. In septic patients <br><br>6. In patients with acute infectious diseases of bacterial origin (dysentery, typhus and paratyphus, tuberculosis) | 1. Under congenital immunodeficiency states <br><br>2. Under acute infectious diseases (influenza, pneumonia) | 1. Autoimmune diseases (disseminated sclerosis, Crohn's disease, systemic *lupus erythematosus*) <br><br>2. Allergic diseases |

The data presented in Table 25 show that Myelopid as an immunocorrecting preparation is most efficient for short term (two to three weeks) transitory immunodeficient states or in patients with various chronic infections accompanied by secondary immunodeficiency. In this case it does not matter what the actual cause of the immunodeficiency was — surgical operation, trauma, continual stress, some chronic pathological process or environmental conditions: the targets for MPs are defects in the immune system.

# Part 2.
# Individual Myelopeptides:
# Structure, Function, Mechanism of Action

As follows from the materials presented in Part 1 of this book, MPs were studied in a strictly logical, step-by-step manner — every stage solved the issues raised from the analysis of data obtained from the preceding experiments. After the discovery of MPs, we comprehensively studied the spectrum of biological activities of these mediators. The results of this work prompted us to develop and apply to clinical practice a new immunomodulator, Myelopid, whose active ingredients are a mixture of MPs isolated from the supernatant of porcine bone marrow cell culture.

At one point, isolation of individual MPs and their structural and functional characteristics became a high priority task. Without solving this problem, it was impossible to understand which peptide was responsible for a particular activity displayed by the mixture of MPs. On the other hand, the synthesis of structurally characterized MPs and their analogs enables one to study molecular mechanisms of action of these new immunoregulators.

The 1990s represented a new stage of utmost importance in resolving the above problem of MPs: several individual peptides were isolated and their amino acid sequences were determined which enabled us to synthesize these MPs and comprehensively study their biological effects and mechanisms of action.

In this part of the book we analyze the results obtained from this stage of our investigation.

## 1. Isolation, Identification and Synthesis of Myelopeptides

For many years the mediators produced by the main organs of the immune system have attracted scientists' attention. However, considerable achievements in the study of immunoregulatory bone marrow peptides, unlike thymus peptides, were made only recently. The researchers studying endogenous bioregulators have to face many problems. The main one is that these compounds are produced in the organism in extremely small amounts; therefore it is necessary to deal with huge amounts of initial material to obtain the final substances in quantities sufficient to determine their structure. Another problem is the formation of many side products of peptide origin in the course of tissue homogenization. So it is not always possible to assert for sure that the isolated substances are real endogenous mediators and not products of unspecific protein degradation.

Modern methods for separation, structural analysis, and chemical synthesis of peptides allows one to characterize in detail peptides produced in the tissues of living organisms. Literature on isolation and identification of bioregulatory peptides shows that, as a rule, most of these compounds have been isolated from homogenized organs or tissues. This is a multistage procedure including reprecipitation, ultrafiltration, various types of low-pressure chromatography (gel chromatography, ion-exchange chromato-graphy) and high-performance liquid chromatography (HPLC) at the final stages.

Though there are many papers describing successful isolation of various endogenous peptide bioregulators, the isolation of each new endogenous peptide bioregulator is a rather difficult task. Two alternative approaches are used to isolate endogenous peptides — total isolation and sequencing of all isolated peptides with subsequent biological testing ("from the structure to the function") or specific isolation of only biologically active compounds based on their activity followed by determination of their structure only ("from the function to the structure"). These approaches have both advantages and disadvantages. As disadvantages of the total isolation method, one can consider the necessity to sequence all isolated peptides, many of which are the products of unspecific protein degradation. Isolation of thymosin-$\beta_1$ and its analogs from fraction 5 of thymosin is a case in point (Goldstein, 1981). The advantage of this approach is that it allows one not to skip biologically active peptides that can be missed because of the complicated testing of a great number of compounds with the use of a wide spectrum of biological test systems, as well as because of inadequate test systems.

Another approach to isolation of biologically active peptides is based on the use of specific biological screening tests. This approach enables the researcher to considerably shorten the time needed to isolate particular peptides; however it is expedient to use this method only for well characterized mixtures containing specific peptides, when adequate test systems capable of revealing definite biological activities are available. A specific example of this directed isolation is the isolation of tuftsin (Bach *et al.*, 1975) and thymus serum factor (Nashioka *et al.*, 1972). Since the bone marrow mediators were comprehensively studied from the functional point of view and several screening model test systems *in vitro* and *in vivo* were developed, we used exactly this approach to isolate individual MPs.

It was demonstrated earlier that MPs display no species specificity. So the supernatant of porcine bone marrow cell culture, that can be available in large quantities, was selected as initial material for MPs' isolation. The cells were cultivated in 199 medium with some additions providing essential conditions maximally approximated to physiological ones to maintain cell vital functions. Hence the cultivation time was rather short (18–20 h), and fetal serum was not added to the culture to prevent protein impurities. The presence of many components of different chemical nature (inorganic salts, amino acids, nucleic bases, vitamins and others) in the culture medium in quantities exceeding the amounts of cell products, significantly hindered the isolation of individual MPs from the supernatant. Methods routinely used at the initial stages of endogenous peptide isolation (gel chromatography, ultrafiltration and others) proved to be inefficient in this case. The problem of isolation of several individual MPs with unknown structure required us to design fundamentally new methodological approaches that enable one to isolate insignificantly small amounts of peptides from a huge amount of the supernatant with minimal loss.

The main problem of the first stage of our work on peptide isolation was standardization of conditions for isolation of the mixture of MPs from the supernatant.

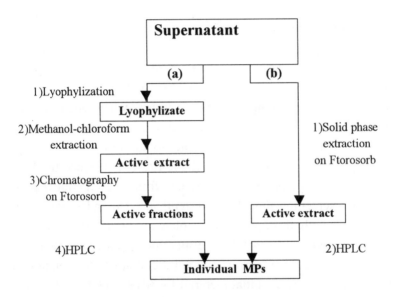

Fig. 28. Scheme for isolation of individual MPs from the supernatant of porcine bone marrow cell culture.

At the beginning of our work, individual MPs were isolated as shown in the scheme (Fig. 28a). To separate inorganic substances, the lyophilized supernatant was extracted with a methanol–chloroform mixture in which the peptides are well soluble as a rule. It enabled us to clear hydrophilic substances from the medium and high molecular weight compounds at a very early stage. The extract obtained, having all types of biological activities characteristic of the initial supernatant was fractionated on a Ftorosorb (macropore glass modified with perfluoropolymer) column. After complete biological testing of Ftorosorb-isolated fractions, only active ones were selected for further fractionation by reversed-phase HPLC (Fig. 29). MP-1 and MP-2 were isolated from one of these fractions and their structures were determined (Phe–Leu–Gly–Phe–Pro–Thr and Leu–Val–Val–Tyr–Pro–Trp, respectively).

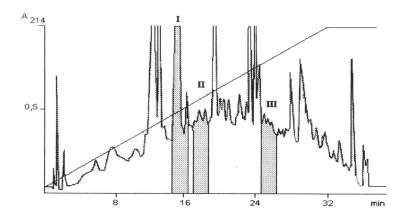

Fig. 29. Separation of the mixture of MPs on an Ultrasphere ODS C18 column (10×250 mm) after initial purification of the supernatant by solid phase extraction. The column was eluted with a 30 min linear gradient of acetonitrile (0–100%) containing 0.1% TFA, flow rate 5.5 ml/min. Active fractions I–III were pooled, concentrated and subjected to an additional separation on rp HPLC. (I) MP-1, MP-2, MP-5, MP-6; (II) MP-4; (III) MP-3.

Assessment of peptide content in the supernatant of various lots (5 lots, 25 l each) demonstrated that the content of individual MPs in the supernatant varies between the lots in a rather broad spectrum (from 10 to 50 pmoles/25 l). In order to obtain individual MPs in quantities sufficient for subsequent amino acid sequencing, we had to process huge amounts of initial supernatant. So in spite of the fact that this scheme enabled us to isolate the first individual MPs, it proved to be unsuitable for further processing of large amounts of the supernatant. It could be explained not only in terms of technical difficulties (the complexity of lyophilization of huge amounts of the supernatant containing inorganic salts in high concentrations) but also in terms of a loss of biologically active peptides due to several freezing and thawing procedures.

Taking this into account, we improved the first stage of MPs isolation and developed a new scheme for isolation of bone marrow peptides (Fig. 28b) based on a solid phase extraction method. According to this method, the initial supernatant passed through a column packed with a sorbent reversibly adsorbing the peptides but not adsorbing inorganic salts and other hydrophilic components of the medium just after cell cultivation.

We selected one such sorbent, Ftorosorb, that when used in fractionation of the lyophilized supernatant (Fig. 28a), enabled us to isolate two individual peptides, MP-1 and MP-2. To simplify the control over isolation of MPs by optical methods, the cells were cultivated in a specially prepared 199 medium not containing phenol red dye. A series of model experiments permitted us to determine an optimal Ftorosorb/supernatant ratio and conditions of separate elution of the mixture of MPs and medium components. By means of reversed-phase HPLC it was established that Ftorosorb washing with 0.1% trifluoroacid (TFA) allowed us to separate the components of 199 medium, while subsequent washing with a 80% acetonitrile solution in 0.1% TFA eluted of a mixture of MPs. According to the data of analytical reversed-phase HPLC, the isolated mixture practically did not contain any medium components. Proteins, lipids and other lipid-soluble compounds that were firmly sorbed on Ftorosorb, could be removed from the column by sorbent regeneration only.

Thus the developed separation scheme allowed us to considerably simplify the whole process and to reduce the time needed for isolation of MPs. The described scheme enabled us to perform several operations at just one stage — that is to concentrate the supernatant, separate the medium components and obtain the biological material suitable for further fractionation by reversed-phase chromatography.

The mixture of MPs obtained after the solid-phase extraction was fractionated on a reversed-phase column (Fig. 28b). All fractions obtained were tested for immunostimulating and differentiating activity. Fractions (I) and (III) with immunostimulating activity and fraction (II) with differentiating activity according to N-terminal amino acid analysis proved to have peptides in quantities sufficient for subsequent sequencing (>200 pM). The peptides of interest were additionally purified by means of analytic reversed-phase HPLC. The quantity of peptides in other active fractions did not exceed 50 pM, so they were not studied further.

The isolated and purified substances of peptide nature were analyzed with the use of a gas-phase sequenator. As a result, two bioactive substances were isolated from fraction (I). Sequencing showed that both are peptides having the following amino acid sequences: Val–Val–Tyr–Pro–Asp and Val–Asp–Pro–Pro. The sequencing of another bioactive substance isolated from fraction (II) allowed us to identify an octapeptide, Phe–Arg–Pro–Arg–Ile–Met–Thr–Pro. The separation of fraction (III) yielded a hexapeptide, Leu–Val–Cys–Tyr–Pro–Gln.

We therefore isolated six new bioactive compounds from the supernatant of bone marrow cell culture, which are:

| | |
|---|---|
| Phe–Leu–Gly–Phe–Pro–Thr | MP-1 |
| Leu–Val–Val–Tyr–Pro–Trp | MP-2 |
| Leu–Val–Cys–Tyr–Pro–Gln | MP-3 |
| Phe–Arg–Pro–Arg–Ile–Met–Thr–Pro | MP-4 |
| Val–Val–Tyr–Pro–Asp | MP-5 |
| Val–Asp–Pro–Pro | MP-6 |

To verify the structure and study the properties of the isolated peptides in detail, they were chemically synthesized by means of a solid-phase method on a PAM polymer using Fmoc/DIPCDI strategy according to a standard program. The synthesized peptides were isolated and purified by means of reversed-phase chromatography. The synthetic peptides demonstrated the same amino acid composition, HPLC patterns and biological activities as the respective native substances.

The structures of the peptides isolated were compared to those of amino acid sequences in the protein–peptide bank of PIR (version 410, 1994). MP-1 and MP-2 are homologous to conservative fragments of α-chain (33–38) and β-chain (31–36) vertebrate hemoglobin, respectively. As to MP-3, MP-4, MP-5, and MP-6, we found no sections of amino acid sequences identical to the structures of proteins registered in the PIR bank. Their precursors are apparently unknown protein molecules.

Hemoglobin fragments corresponding to amino acid sequences of MP-1 and MP-2 peptides are not restricted by base amino acid pairs in the protein chain, so they cannot be products of routine processing. We can assume that these peptides are products of proteolysis of some unknown precursor protein or products of nonspecific hemoglobin degradation. The possible role of hemoglobin as a precursor protein for a number of biologically active peptides isolated from bone marrow and brain homogenates of various animals is substantiated by the data obtained by V.T. Ivanov and co-workers (Ivanov *et al.*, 1997).

## 2. MP-1 is a Myelopeptide with Immunocorrective Activity

We studied the immunoregulatory properties of MP-1 in various *in vivo* and *in vitro* models by means of experimental techniques that enabled us to reveal its ability to immunocorrection, as well as to clarify some details in the mechanism of its action (Mikhailova *et al.*, 1994). Having this peptide in a homogenous form obtained by chemical synthesis, we could investigate its immunoregulatory properties using a standard substance and reveal some dose-dependent regularities. First of all, we studied the ability of MP-1 to stimulate antibody formation *in vitro* in the productive phase of the immune response to SRBC. This was the model in which we had discovered immunoregulatory mediators of the bone marrow, MPs, for the first time (Petrov and Mikhailova, 1972; Petrov *et al.*, 1975).

### 2.1. MP-1 Stimulates Antibody Formation at the Peak of Immune Response

Mice were immunized with a 5% SRBC suspension injected subcutaneously in a volume of 0.1 ml into all four footpads. The second immunization was conducted after two to three weeks with the same antigen at the same dose. On the fourth day after the secondary immunization, the mice were decapitated, popliteal and groin lymph nodes were isolated. The lymph node cells at a concentration of $2 \times 10^6$ cells/ml were incubated with and without the peptide for 18-20 h at 37°C in 199 culture medium with the addition of L-glutamine (2 mM), HEPES buffer (10 mM), penicillin (10 units/ml) and 10% fetal calf serum. On completing the incubation, the number of AFC was determined in every culture. MP-2 was used in these experiments as one of the controls. The peptides were added to the plate wells at a concentration range of $5 \times 10^{-4}$–$5 \times 10^2$ pM. The effect of antibody stimulation was assayed by comparing the AFC number in the cultures with and without the peptides. The results are shown in Fig. 30. In distinction to MP-2, MP-1 produces a dose-dependent, statistically significant effect of stimulation of antibody formation *in vitro* at concentrations of 0.005 – 500 pM. The antibody-stimulating effect of MP-1 reaches approximately 150% at concentrations of

0.5 – 50 pM (Fig. 30, curve 1). MP-2 had no effect on the level of antibody production in this experimental model at concentrations used (Fig. 30, curve 2).

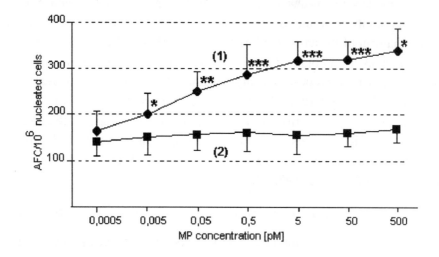

Fig. 30. Effect of MP-1 (curve 1) and MP-2 (curve 2) on the number of AFC in the culture of immune lymph node cells. *p<0.05, **p<0.01 and ***p<0.001 vs. control.

Thus, one of the immunoregulatory properties of MP-1 is its ability to increase the number of AFC at the peak of the immune response.

An increase in the number of AFC in the population of mature antibody producers under the influence of MP-1 could be a consequence of cell proliferation enhancement. So we studied the mitogenic effect of MP-1.

Mouse spleen cells were incubated with MP-1 ($10^{-9}$–$10^{-6}$ g/ml), Con A (2 μg/ml) or LPS (50 μg/ml) for 72 h. [$^3$H] thymidine (0.5 Ci/well) was added 6 h before the completion of cell cultivation. The assay of [$^3$H] thymidine incorporation into DNA showed that in distinction to Con A and LPS — which are strong mitogens — MP-1 at concentrations used did not enhance DNA replication (Table 26). Hence, the increase in the number of AFC in the population of mature antibody producers under the influence of MP-1 occurs without cell-division and this is apparently connected with the transition of so-called "silent" cells from the rest state to active antibody production (Petrov and Mikhailova, 1972). We suppose this transition to be controlled by helper–suppressor interactions which MP-1 is able to affect.

Table 26. [$^3$H] thymidine incorporation in mouse spleen cells stimulated with MP-1 or mitogens.

| Stimulus | Dose, µg/ml | Cpm |
|----------|-------------|-----|
| Control | – | 284±54 |
| Con A | 2 | 32551±3145 |
| LPS | 50 | 3856±527 |
| MP-1 | 0.001 | 289±78 |
| | 0.01 | 267±66 |
| | 0.1 | 311±85 |
| | 1 | 296±61 |

Previously it was demonstrated that the antibody-stimulating effect of the native mixture of MPs is connected with their ability to abolish the suppressive activity of T lymphocytes (Mikhailova et al., 1976). We decided to study the effect of MP-1 both on the induction and manifestation of the functional activity of T suppressors (Petrov et al., 1994).

The suppressive activity of T lymphocytes was induced in the suspension of mouse spleen cells by cultivating them with Con A (Peavy and Pierce, 1974). After 48 h incubation, the washed cells were mixed with immune lymph node cells at a ratio of 1:1, incubated for 18-20 h and the number of AFC was determined in the mono- and mixed cultures. Figure 31 shows that mouse spleen cells incubated with Con A decreased the antibody production almost by two times in the immune lymph node cells because of the induction of T suppressors (Peavy and Pierce, 1974).

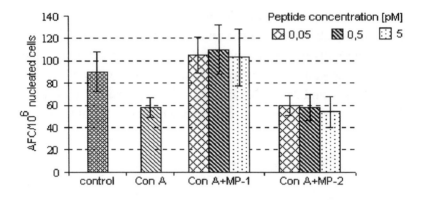

Fig. 31. Effect of MP-1 and MP-2 on the induction of suppressor cells in the suspension of Con A-activated mouse spleen cells. Suppressor cells were detected by decrease of the antibody production in the culture of immune lymph node cells.

The addition of MP-1 to the spleen cells together with Con A prevented induction of T suppressor cells. Mouse spleen cells incubated with Con A in the presence of MP-1 did not diminish the antibody production in the culture of immune lymph node cells. At all three concentrations used, MP-1 abolished the induction of the suppressive activity of T lymphocytes. In distinction to MP-1, MP-2 did not prevent the induction of T suppressors by Con A. Mouse spleen cells incubated with Con A in the presence of MP-2 suppressed the antibody formation in the population of immune lymph node cells like spleen cells incubated with Con A without any peptides (Fig. 31).

The influence of MP-1 on the functional activity of T suppressors was studied by experiments in which peptides (MP-1 or MP-2) were added to mixed cultures of immune lymph node cells and Con A-induced T suppressors. After 45 min incubation the change in the number of AFC was determined. Figure 32 demonstrates that the joint cultivation of immune lymph node cells with T suppressors resulted in a twofold decrease in the number of AFC.

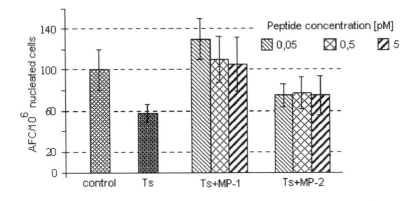

Fig. 32. Effect of MP-1 and MP-2 on the functional activity of T suppressors.

The presence of MP-1 in this mixture prevented the inhibitory effect of T suppressors on antibody formation. As to MP-2, this peptide abolished the inhibitory effect of T suppressors only partly. The number of AFC in the mixed culture of immune lymph node cells and T suppressors in the presence of MP-2 decreased only to 78%, whereas in the culture containing T suppressors in the absence of MP-2 this decrease reached 60%.

The data presented indicate that the increase in the number of AFC in the population of mature antibody producers under the influence of MP-1 occurs apparently because of its interference with helper–suppressor interactions. As a result, the suppressive activity of T lymphocytes is decreased or completely abolished.

The ability of MP-1 to control the function of regulatory T cell subsets characterizes this peptide as a bioregulator participating in the development of a complex network of molecular and cellular events that provides a normal level of the immune response. It is known that these subpopulations and T cell suppression, in particular, play a significant role in T lymphocyte control over the level of B cell response to an antigen (Green and Flood, 1983).

## 2.2. MP-1 Ability of Immunocorrection

The supposed mechanism of immunoregulatory action of MP-1 at the level of mature antibody producers evidently underlies the immunocorrective effect of this peptide, because many immunodeficiency states are accompanied by helper/suppressor disbalance.

We used some immunodeficiency models of various etiology to reveal and study the immunocorrective ability of MP-1.

One of them were mice irradiated with γ-rays from $Co^{60}$ at a dose of 2 Gy. 12 days after irradiation the mice were immunized with SRBC and the number of AFC in the spleen was determined on the fifth day after the immunization. Some groups of the irradiated animals were injected with various doses of MP-1 or MP-2 on days 0, 1, 2, 3 after the immunization. The results of the experiments are shown in Fig. 33.

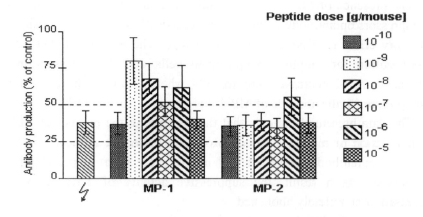

Fig. 33. Effect of MP-1 and MP-2 on the antibody production in irradiated mice. Each bar indicates the number of AFC as percentage to the control.

The irradiation of mice resulted in a decrease in AFC number in the spleen to 36.8%. MP-1 injection revealed this by augmenting the immune response. The maximal effect was observed at a dose of $10^{-9}$ g/mouse. Doses of $10^{-6}$, $10^{-7}$, $10^{-8}$ g/mouse proved to be less efficient, though the increase in the AFC number at these doses was statistically significant as well ($p<0.05$). MP-1 doses of $10^{-5}$ and $10^{-10}$ did not cause increase in antibody production in the irradiated mice.

The injection of MP-2 instead of MP-1 at the same doses in the irradiated mice did not result in augmentation of antibody production. Only one dose of MP-2, $10^{-6}$ g/mouse, caused a slight but statistically significant increase in antibody production ($p<0.05$).

Thus, MP-1 displays a pronounced immunocorrective effect in irradiated animals by enhancing the decreased level of antibody production up to 80% of the level in non-irradiated control animals.

Another immunodeficiency model was mice treated with cytostatic cyclophosphamide (Cy). The mice were inoculated with Cy at a dose of 200 mg/kg. One or nine days later, the mice were immunized with SRBC for an antibody response assay. Some animals were also injected with MP-1 or MP-2 at doses of $10^{-6}$ and $10^{-7}$ g/mouse on days 0, 1, 2, 3 after the immunization. On the fifth day of the immune response we determined the number of AFC in the spleen of mice in the control and experimental (treated with MP-1 or MP-2) groups.

The results presented in Figure 34 show that immunization of the animals on the next day after Cy treatment resulted in a practically complete inhibition of antibody production. MP-1 or MP-2 inoculation in this time had no influence on the number of AFC in the spleen. Under immunization on the ninth day after Cy treatment, the number of AFC in mouse spleen increased significantly as compared to that on the first day and reached 48.7% of the control. MP-1 injection at this stage resulted in a considerable increase in AFC number, and at a dose of $10^{-6}$ g/mouse it reached a level comparable to that of control animals. MP-2 at all doses used showed no effect on the number of AFC in the spleen of Cy treated mice.

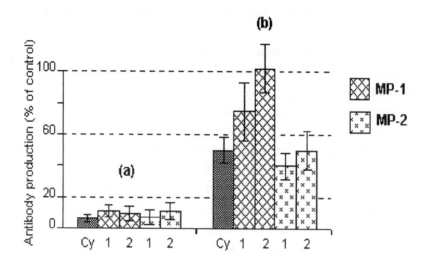

Fig. 34. Effect of MP-1 and MP-2 on the number of AFC in the spleen of Cy-treated mice immunized with SRBC on the next day after Cy treatment (a) or nine days (b) after it. Cy, Cy-treated mice; 1, 2, doses of MPs ($10^{-7}$ and $10^{-6}$ g/ml, respectively).

In the next series of experiments we studied the effect of MP-1 on antibody production under the conditions of T or B cell deficiency that was induced in the following models. The mice were injected intravenously with Con A (T cell mitogen) or LPS (B cell mitogen). After two days they were inoculated with Cy. It is known that Cy mainly influences the actively proliferating cells. So we used mice with a lower level of T cells (Con A-treated mice) or B cells (LPS-treated mice).

Figure 35-1 shows the data on MP-1 influence on antibody production under the conditions of T cell deficiency. Cy injection into mice previously treated with Con A resulted in a significant decrease in the immune response to SRBC (37.3% of the control level) as compared to the animals treated with Cy (57.7%) or Con A (78.2%) only. The injection of MP-1 at a dose of

$10^{-6}$ g/mouse to the animals successively treated with T cell mitogen and Cy did not increase the number of AFC in the spleens.

The combined inoculation in mice of LPS and Cy induced B cell deficiency that resulted in a statistically significant decrease in the number of AFC in mouse spleens (40.1% of the control level). The mice treated with either LPS or Cy showed less decrease in the number of AFC (74.4% and 64.1%, respectively) (Fig. 35-2). The injection of MP-1 at a dose of $10^{-6}$ g/mouse to B cell deficient animals resulted in a slight increase in the number of AFC in the spleen (up to 53.3%). The results of this experiment show that the presence of T cells is necessary for manifestation of the MP-1 antibody stimulating effect.

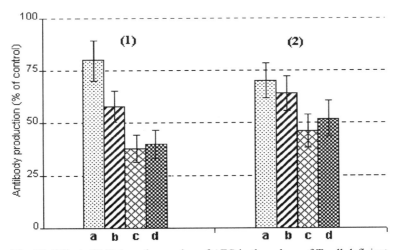

Fig. 35. Effect of MP-1 on the number of AFC in the spleen of T cell deficient (1) and B cell deficient (2) mice induced by Cy. Mice treatment: (1a) Con A; (1b) Cy; (1c) Con A + Cy; (1d) Con A + Cy + MP-1; (2a) LPS; (2b) Cy; (2c) LPS + Cy; (2d) LPS + CY + MP-1.

To study the influence of MP-1 on the mitogen-induced proliferation in spleen cells of Cy treated mice, MP-1 or MP-2 were injected on days 1, 2, 3 after Cy treatment. The peptide dose used was $10^{-6}$ g/mouse. On the fourth day, the spleen cells of these mice were cultured in the presence of Con A (2 μg/ml) or LPS (50 μg/ml) for 72 h. The cultures were pulsed with 1 mCi of [$^3$H] thymidine per well for the last 6 h. The cells were harvested and [$^3$H] thymidine incorporation was measured. As one can see from Fig. 36, the proliferative response of spleen cells to Con A was 30.1% and that to LPS was 8.6% of the level of spleen cells of normal animals. MP-1 inoculation resulted in augmentation of spleen cell proliferative response to Con A (47.2% vs 30.1% in the control; $p < 0.05$) and had no effect on the LPS-induced proliferative response. In distinction to MP-1, MP-2 had no influence either on Con A-, or LPS-induced proliferation of spleen cells of Cy treated mice.

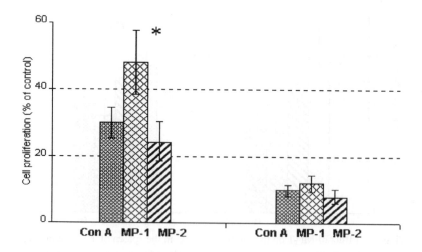

Fig. 36. Effect of MP-1 and MP-2 on the proliferation of Cy-treated mouse spleen cells *in vivo*. *$p < 0.05$ vs. Con A-stimulated control.

The data presented suggest that in distinction to MP-2, MP-1 has an immunocorrective effect on the functional activity of lymphoid cells in immunodeficient animals: it increases the level of antibody production in irradiated mice (Fig. 33), restores the humoral immune response in Cy-treated mice (Fig. 34), and enhances the proliferative response of spleen cells in immunodeficient animals to the T cell mitogen (Fig. 36). The depressive influence of irradiation or Cy on antibody production in response to SRBC is well known at the present time. The dynamics of restoration of the thymus-dependent immune response in irradiated and Cy-treated mice, have similar regularities. The following stages can be distinguished here: the stage of a sharp decrease in antibody production just after the treatment, the latent stage and the recovery stage (Anokhin and Yarilin, 1984). The data presented in Fig. 34 show that MP-1 is ineffective just after the immunodepressive treatment, that is in the period when mass death of lymphocytes occurs and the functions of live cells are impaired. The maximal efficiency of MP-1 is observed in the recovery period of the thymus-dependent antibody response (12 days after the irradiation of the animals and 9 days after Cy treatment) and is apparently explained by the fact that by this time the lymphoid organs begin to be occupied with normal lymphocytes, including target cells for MP-1 capable of cooperative interaction in the immune response.

The directed induction of deficiency in mice *in vivo* due to selective elimination of either T or B cells by means of Cy showed that the MP-1 immunocorrective effect takes place via the T cell population. The successive injection of Con A and Cy to mice caused a T cell anergy resulting in a sharp decrease in the response to the thymus-dependent antigen. MP-1 injection to T cell deficient mice did not cause any significant changes in antibody production in the spleen in the response to SRBC. However, in mice with B cell deficiency induced by successive LPS and Cy injections, MP-1 caused an increase of AFC number in the spleen in response to SRBC. This increase in antibody production in response to the thymus-dependent antigen under the influence of MP-1 is apparently mediated by T cells that have preserved their functional activity, thus promoting recruitment of undamaged B cells into the differentiation process. The data showing that MP-1 administration to Cy-treated mice caused an increase in Con A-induced spleen cell proliferation but did not influence the LPS-induced proliferative response of spleen cells confirm that MP-1 immunoregulatory activity occurs via T lymphocytes.

## 2.3. MP-1 Specific Binding with Target Cells

The availability of a structurally characterized and synthesized MP-1 enabled us to analyze the possibility of its binding to the surface of lymphoid cells and to determine the target cell for it using a cytofluorimetry method (Mikhailova *et al.*, 1994).

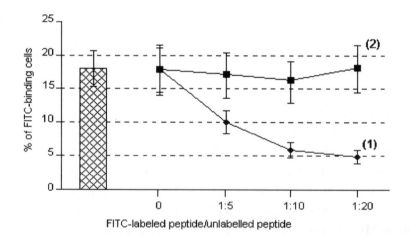

Fig. 37. Binding of FITC-labeled MP-1 to target cells in mouse spleen cell suspension. The bar indicates the initial percentage of FITC-positive cells in the mouse spleen cell suspension. (1) FITC-labeled MP-1/unlabeled MP-1; (2) FITC-labeled MP-1/unlabeled MP-2.

FITC-labeled MP-1 was incubated with a suspension of mouse spleen cells for 30 min at 4°C. After incubation the cells were washed twice and analyzed on an EPICS-5 flow cytofluorimeter. On average 10,000 cells were analyzed in every specimen and the percentage of cells bound with the labeled peptide was determined.

The data presented in Fig. 37 show that there are 18% FITC-positive cells in the mouse spleen cell population. The binding of MP-1 to the cell surface is specific since the addition of an excess amount of unlabeled MP-1 at various concentrations to the incubation mixture results in a dose-dependent loss of the label from the binding sites on the cell surface (Fig. 37, curve 1). Another unlabeled peptide with a different amino acid sequence, MP-2, added at the same concentrations did not displace FITC-labeled MP-1 from the binding sites on the cell membrane (Fig. 37, curve 2). The Kd was determined by the Scatchard method (Scatchard, 1949), and proved to be $1.2 \times 10^{-7}$ M which indicated a low affinity for the MP-1 receptor.

To determine which cell type whithin the heterogeneous mouse spleen cell population binds with MP-1, we used the method of double fluorescent staining. Mouse spleen cells were treated with RhITC-labeled MP-1 and FITC-labeled monoclonal antibodies to surface markers of T lymphocytes (anti-Thy 1.2) or B lymphocytes (anti-Ig). The binding of labeled MP-1 with T or B lymphocytes was measured in the flow cytofluorimeter. The specificity of binding was assessed by adding excessive amounts of unlabeled MP-1 to the incubation mixtures. The results of the experiments are presented in Fig. 38.

The specific binding of the RhITC-labeled MP-1 occurs with T lymphocytes only (18%); unlabeled MP-1 displays RhITC-labeled MP-1 from the binding sites on the T cell surface in a dose-dependent manner. The binding of MP-1 with B lymphocytes (5%) is nonspecific since no displacement of the label by excessive amount of unlabeled MP-1 was be observed.

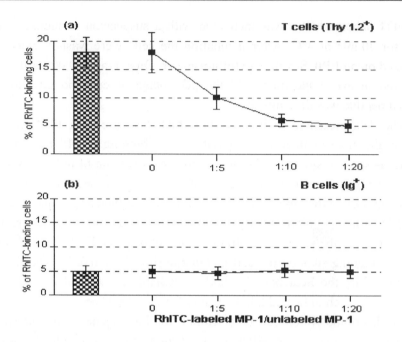

Fig. 38. Binding of RhITC-labeled MP-1 with (a) T lymphocytes (FITC-labeled anti Thy 1.2 antibodies) or (b) B lymphocytes (FITC-labeled anti-IgG antibodies). The bars indicate the initial percentage of RhITC-positive cells in the appropriate cell population.

Using labeled monoclonal antibodies to the markers of regulatory T lymphocyte subsets, namely helpers (anti-L3T4 for CD4+cells) and suppressors (anti-Lyt-2 for CD8+ cells), we answered the following question: which type of T lymphocyte subpopulations has a specific receptor for MP-1? The data presented in Fig. 39 show, that the target cells for MP-1 are CD4+ lymphocytes, that is, T helpers. 16% CD4+ cells were RhITC-positive, and this binding was specific; while only 6% CD8+ cells unspecifically bound with the RhITC-labeled MP-1.

In all experiments using either a mixture of nonidentified MPs or a synthesized MP-1, we observed the absence of an inhibitory effect of T suppressors on antibody production in the presence of this peptide. Thus a question arises, how does MP-1 abolish the suppressive activity of T lymphocytes having T helper as a target cells.

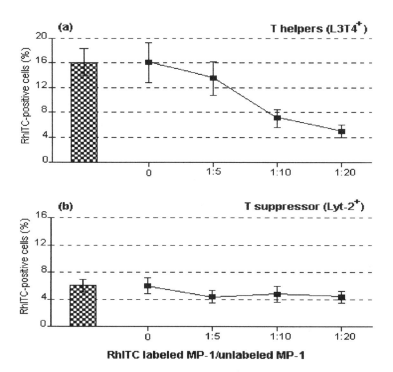

Fig. 39. Binding of RhITC-labeled MP-1 with T lymphocyte subsets, (a) T helpers (FITC-labeled anti L3T4 antibodies) or (b) T suppressors (FITC-labeled ant- Lyt2 antibodies). The bars indicate the initial percentage of RhITC-positive cells in the appropriate cell subsets.

We conducted experiments on mouse spleen cells with the CD4+/CD8+ ratio shifted towards CD8+ cells. The mice were injected with Con A at a dose of 100 µg/mouse. After 48 h T suppressors accumulated in the spleen of these animals. The addition of these cells to lymph node cells obtained from mice at the peak of the immune response to SRBC resulted in a 2-fold decrease in the number of AFC (Table 27, line 2). When a portion of spleen cells from Con A-treated animals which had been previously incubated for 45 min with 5 pM MP-1 was added to the immune lymph node cells, the level of antibody production in the population of mature AFC did not decrease further, but increased above even the basal level (Table 27, line 3).

Table 27. Effect of MP-1 on the functional activity of Con A-induced T suppressors.

| Cell mixture | $AFC/10^6$ nucleated cells | Suppression index | p |
|---|---|---|---|
| Immune lymph node cells (control) | 97.8±15.3 | | |
| Immune lymph node cells + $T_s$ | 58.8±11.6 | 0.6 | 0.001 |
| Immune lymph node cells + ($T_s$ + MP-1) | 130.9±27.2 | 1.34 | 0.01 |

The cytofluorimetric analysis of spleen cells from Con A-treated mice before and after incubation with MP-1 are presented in Fig. 40. One can see that if the CD8+/CD4+ (Ts/Th) ratio of cells bound with MP-1 in the spleen of normal mice is 1:3, then in the spleen of Con A-treated animals this ratio was sharply shifted towards suppressor cells and became 2.5:3.

The binding of CD8+ cells with RhITC–MP-1 increased almost 2.5-fold as compared to the control group. As is shown above, this binding is unspecific (Fig. 39). The increase in the number of T suppressors unspecifically bound with RhITC–MP-1 apparently occurs due to the increase in the absolute number of CD8+ cells in the spleen of Con A-treated mice. The number of CD4+ cells is fixed, so their specific binding with RhITC–MP-1 is the same as in the control group (Fig. 40).

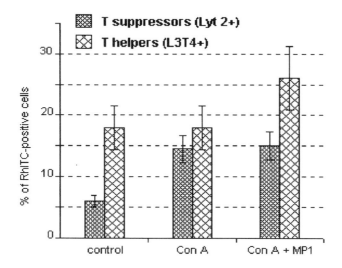

Fig. 40. Binding of RhITC-labeled MP-1 with T helpers (FITC-labeled anti-L3T4 antibodies) and T suppressors (FITC-labeled anti-Lyt2 antibodies) in mouse spleen cell suspension in control and Con A-induced T suppression.

The CD8+/CD4+ disbalance in mouse spleen under the influence of Con A partly normalized after incubation of splenocytes with MP-1. The CD8+/CD4+ ratio was 1.7:3 as compared to 1:3 in the control group. The number of CD8+ cells remained at the same level (15%), while the number of CD4+ cells that bound MP-1 increased up to 26%. Their binding with RhITC-labeled MP-1 was specific.The question which molecular and cellular mechanisms underlie the enhancement of MP-1 binding with the helper subpopulation of T lymphocytes under CD8+/CD4+ disbalance is still not resolved. It is quite possible that in the spleen of Con A-treated mice a receptor to this regulatory peptide is expressed on the additional CD4+ cells under the influence of MP-1. This question as well as the role of Th-1 and Th-2 in the regulatory process will be studied in further experiments.

As to the present data, we can postulate that the correction of antibody production under the influence of MP-1 in conditions of CD8+/CD4+ disbalance occurs at the expense of the binding of this regulator to the receptor on the CD4+ cell surface, thus causing their activation and recovery of the balance with regard to the enhanced suppressor activity of T lymphocytes.

The data on the biological activity of MP-1 and on the mechanisms of its manifestation obtained so far enables us to assert with highest probability that MP-1 is an endogenous immunoregulatory peptide participating in the bioregulation of immune homeostasis. Selective binding of MP-1 with a specific receptor on the cell surface of one of the regulatory T lymphocyte subpopulations under their impaired balance, manifestation of the correcting effect of MP-1 both *in vitro* and *in vivo*—all these data confirm the physiological value of this peptide regulator and its role in the normal functioning of the immune system.

## 3. MP-2 is a Myelopeptide with Antitumor Effect

It is known that the immune system and T lymphocytes, in particular, play a key role in the defense mechanisms of an organism against the development and spreading of malignant neoplasms (Miescher *et al.*, 1986; Weil-Hillman *et al.*, 1991; Renner and Pfreundschuh, 1995). As a rule, tumor growth is accompanied by a decrease in body resistance and by considerable immune disorders, mainly in a lowered function of the T system of immunity. It was shown that tumor cells produce some substances that suppress T lymphocyte function as tested by a T cell proliferative response to the mitogen (Chiao *et al.*, 1986; Miescher *et al.*, 1986). Conditioned medium from the human myeloleukemia cell line HL-60 (HL-60 CM) inhibited the proliferative response of human peripheral blood T cells to PHA. This T lymphocyte dysfunction was accompanied by a drastically reduced IL-2 production (Chiao *et al.*, 1986).

Previous investigations on functional activities of the mixture of MPs revealed the ability of these mediators to restore the T lymphocyte function suppressed by tumor toxins (Strelkov and Mikhailova, 1989). Hence this experimental model was used along with some others to test the biological activity of synthesized individual MPs.

### 3.1. Myelopeptide 2 Recovers T lymphocyte Functional Activity Inhibited by Tumor Cells

Among all tested individual MPs, only MP-2 displayed the ability to abolish suppression of the T lymphocyte proliferative response to PHA induced by addition of conditioned medium (CM) from HL-60 (Strelkov *et al.*, 1996). It was revealed in the following experiments.

The human myeloleukemia cell line HL-60 was maintained in a standard medium (RPMI-1640) with addition of 15% fetal calf serum, 20 mM HEPES buffer, 2 mM L-glutamine and 50 µg/ml gentamycin. Initially the cells were cultivated at a concentration of $2 \times 10^5$ cells/ml and HL-60 CM was collected during the logarithmic growth phase (the third/fourth day) and further used as tumor cell products (Chiao *et al.*, 1986; Miescher *et al.*, 1986). T lymphocytes from the peripheral blood of healthy donors ($1 \times 10^6$ cells/ml) were cultivated in the standard medium for 72 h in the presence of 3 μg/ml PHA. All cultures with the exception of controls contained 10% (v/v) HL-60 CM as the suppressive substance. 4 h before terminating the cultivation, each culture was labeled with [$^3$H]-thymidine (2 μCi/ml). The cells were harvested and their DNA radioactivity was measured. The mean counts per min (cpm) in triplicate cultures were assayed. The results of the experiments are presented in Table 28.

Table 28. MP-2 abolishes the inhibition of T lymphocyte proliferative response to PHA induced by HL-60 CM.

| Substances added to T lymphocytes | [$^3$H]-thymidine incorporation | |
|---|---|---|
| | *cpm × $10^3$* | *% of control* |
| – | 0.3±0.1 | 1.4±0.7 |
| PHA (control) | 22±4.2 | 100.0±19.1 |
| PHA + HL-60 CM | 10±2.8 | 45.4±12.7 |
| PHA + HL-60 CM + MP-2 | | |
| 100 μg/ml | 20.8±3.1 | 94.5±14.1 |
| 50 μg/ml | 17.0±2.4 | 77.3±10.9 |
| 5 μg/ml | 13.0±2.0 | 59.1±9.1 |
| 1 μg/ml | 12.6±1.8 | 57.3±8.2 |
| 0.1 μg/ml | 10.4±1.7 | 47.2±8.4 |
| PHA + HL-60 CM + MP-1 | | |
| 100 μg/ml | 6.0±1.0 | 27.3±4.5 |
| 50 μg/ml | 8.0±1.4 | 36.4±6.3 |
| 5 μg/ml | 10.2±2.6 | 45.4±11.8 |

T lymphocyte proliferative response to PHA was reduced down to 50% of the control value under the influence of HL-60 CM. MP-2 abolished the HL-60 CM inhibitory effect on the T lymphocyte response to the mitogen.

The recovery of PHA-induced T lymphocyte proliferation under the influence of MP-2 was dose-dependent. MP-2 at concentrations of 0.1-1.0 µg/ml had minimal recovery effect, while at concentrations of 50 and 100 µg/ml it induced a 80-90% recovery of the T lymphocyte proliferative response to the mitogen. Another peptide, MP-1, did not display such an effect on the inhibited T lymphocyte function under similar experimental conditions. At concentrations of 50 and 100 µg/ml it even enhanced the suppressive action of tumor cell products on PHA-induced T cell proliferation (Table 28).

To answer the question of whether MP-2 prevents the suppressive effect of tumor toxins on T lymphocyte function or restores the functional activity of impaired T cells, we conducted another series of experiments. MP-2 was added to PHA-stimulated T lymphocytes simultaneously, 1 h or 24 h after their exposure to HL-60 CM. The results presented in Table 29 show that the addition of MP-2 to PHA-stimulated T lymphocytes 1 h after the suppressive agent (HL-60 CM) even slightly increased the T lymphocyte response to PHA (114% of the control; $p < 0.05$). MP-2 addition after 24 h did not restore the proliferative response of T lymphocytes completely (80.3% of the control) but accounted for a 30% increase in the response as compared to that without MP-2. These data show that MP-2 can restore T lymphocyte function which was damaged by tumor cell products.

Table 29. MP-2 restores the PHA response of T lymphocytes when added to T cells after HL-60 CM.

| Experimental conditions | T lymphocyte response to PHA, % of the control |
|---|---|
| PHA (control) | 100 |
| PHA + HL-60 CM | 51.3±6.8 |
| PHA + HL-60 CM + MP-2 (100 µg/ml) | |
| added simultaneously with HL-60 CM | 87.1±4.3 |
| added 1 h after HL-60 CM | 114.3±6.5 |
| added 24 h after HL-60 CM | 80.3±4.8 |

Using flow cytometry analysis, we managed to show that the suppression of T lymphocyte function under the influence of tumor cell products and its further restoration in the presence of MP-2, is accompanied by changes in cell phenotype. T lymphocytes from human peripheral blood were incubated with mouse anti-CD3, CD4 or CD8 monoclonal antibodies, then with FITC-labeled rabbit IgG F(ab')$_2$ fragments against mouse IgG (H+L). After washing and fixation, the cells were scanned in the flow cytofluorimeter (EPICS-ELITE Counter, USA).

Table 30. Effect of MP-2 and IL-2 on the human T lymphocyte phenotype damaged by the products of leukemia HL-60 cells.

| Substances added to T lymphocytes | Phenotypic markers: % of positive cells and relative fluorescence intensity (in brackets) | | |
|---|---|---|---|
| | CD3 | CD4 | CD8 |
| – | 79.3±2.8 (100) | 50.4±3.2 (100) | 29.1±2.2 (100) |
| HL-60 CM | 70.6±2.5 (75) | 41.5±29 (78) | 30.0±21 (99) |
| HL-60 CM + MP-2 (100 µg/ml) | 79.4±2.9 (99) | 49.2±3.0 (98) | 29.4±2.3 (100) |
| HL-60 CM + IL-2 (?/ml) | 77.2±2.6 (68) | 42.6±3.1 (76) | 36.3±2.1 (134) |

The results of the analysis presented in Table 30 show that the addition of HL-60 CM in the culture of human peripheral blood T lymphocytes reduces the proportion of CD3+ and CD4+ cells. Simultaneously with the decrease in the quantity of these cells in the whole T cell population, the density of these antigens on the cell surface (relative values of fluorescence intensity) was also reduced. MP-2 restores these phenotypic disorders. At the same time, the expression of CD8 antigen did not change either under HL-60 CM or under MP-2 influence; however it increased under IL-2 exposure.

So, the leukemia cell products change the expression of CD3 and CD4 antigens on T lymphocytes, thus enhancing the level of "defective" T lymphocytes and, consequently, suppressing the T lymphocyte function.

By restoring the defects induced by tumor toxins, MP-2 facilitates normal T lymphocyte function. The recovery of the damaged phenotype in CD4+ T cells under the influence of MP-2 is clearly demonstrated by the histogram presented in Fig. 41. We can see the changes in the fluorescence of CD4+ T lymphocytes caused by HL-60 CM. The main CD4 fluorescence peak is reduced as compared to the control. And what is more, HL-60 CM generates a "defective" T lymphocyte subpopulation with a reduced density of CD4 antigen on their surface. These data together with the results presented in Table 30 show the changes in T lymphocyte phenotype caused by tumor cell products. MP-2 corrects this and restores the normal functioning of the T system of immunity.

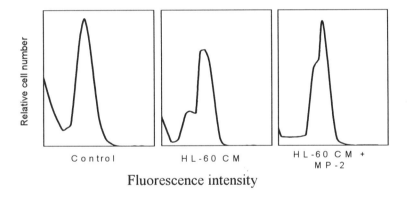

Fig. 41. Effect of MP-2 on CD4 antigen expression in human peripheral blood T lymphocytes.

Since CD4+ T lymphocytes are responsible for both IL-2 production and its receptors (Kuo and Robb, 1986), it is highly probable that MP-2 restores the T cell proliferative response to PHA by influencing the IL-2/IL-2R system. One of the causes for the decrease in T lymphocyte proliferative response to the mitogen under the influence of tumor cells can be the reduced ability of T cells to produce IL-2 and/or to express IL-2R. This assumption is substantiated by the restoration of mitogen-induced T lymphocyte proliferation in this experimental model by the addition of exogenous IL-2 (Chiao *et al.*, 1986) as well as by data on the reduced CD25+ cell level (these are IL-2R-bearing cells) caused by HL-60 CM, and its normalization in the response to MP-2 or IL-2 addition (Strelkov *et al.*, 1996). However, the recovery mechanisms of MP-2 and IL-2 action on the PHA-response of T lymphocytes are apparently different. The data presented in Table 30 show that unlike MP-2, which promotes recovery of CD3 and CD4 antigen expression, IL-2 enhances CD8 antigen expression and reduces the density of CD3 antigen on T cells in the presence of HL-60 CM (mean fluorescence intensity reduced up to 68.0 as compared to 75.0 in the control). At the same time, IL-2 normalized the level of CD3+ but not that of CD4+ cells.

These results suggest that MP-2 recovery of CD3 and CD4 antigen expression impaired by HL-60 CM, alsonormalizes the IL-2/IL-2R , thereby promoting the T lymphocyte response to mitogen. IL-2 on the other hand enhances the number of PHA target cells by decreasing the CD3 antigen density on T cells and generating active CD8+ T lymphocytes. It is known that PHA stimulates the proliferation of mainly CD8+, but not that of CD4+ T lymphocytes (Adriaansen *et al.*, 1990).

The revealed ability of MP-2 to abolish the inhibitory effect of tumor cells through the functional activity of T lymphocytes *in vitro* suggests the possibility of an antitumor effect of this peptide *in vivo*. The efficient immune defense in a tumor-bearing organism must promote the inhibition of tumor growth and active elimination of tumor cells. This very principle forms the basis of immunotherapy methods used in oncology nowadays. In this respect we studied the antitumor effect of MP-2 *in vivo* using various models of implanted and spontaneous tumors in mice.

### 3.2. Antitumor Action of Myelopeptide 2 in vivo

The antitumor effect of MP-2 *in vivo* was assessed in various implanted tumors: lympholeukemia P-388, Lewis lung adenocarcinoma,, mammary adenocarcinoma Ca-755, melanoma B-16, sarcoma S-180, and in spontaneous mammary tumor in BLDR-Rb(8.17)1Iem mice which are known to have such mammary tumors after confinement in 100% cases (Moiseeva *et al.*, 1991).

The experiments on implanted mouse tumors were conducted jointly with Prof. G.K. Gerasimova and Dr. E.M. Treshchalina (Cancer Research Center, Russian Academy of Medical Sciences, Moscow). The tumors were transplanted to mice and on the third day after the implantation, MP-2 injections were started. The most suitable and rather efficient scheme for administration of the peptide was chosen from five schemes studied — MP-2 at a dose of 1 or 2 mg/kg was injected twice with a 96 h interval between the injections. The tumor size was measured after certain intervals after the transplantation. Table 31 shows the results of the experiments conducted. MP-2 displays a pronounced inhibitory effect on the growth of all tumors studied. The level of tumor growth inhibition under MP-2 influence reached 70-80%.

Table 31. MP-2 effect on the growth of various transplantable mouse tumors.

| Tumor | Days after tumor transplantation | | | | | | | |
|---|---|---|---|---|---|---|---|---|
| | 7 | | | 14 | | | 18 | |
| | Tumor size, $mm^3$ | | $V_e/V_c$ (%) | Tumor size $mm^3$ | | $V_e/V_c$ (%) | Tumor size, $mm^3$ | | $V_e/V_c$ (%) |
| | control | experim. | | control | experim. | | control | experim. | |
| Lympholeukemia P-388 | 364±131 n=10 | 201±90* n=10 | 55 | 2190±386 n=10 | 778±247** n=10 | 35 | 6380±403 n=10 | 2268±361*** n=10 | 35 |
| Breast adenocar-cinoma Ca755 | 949±176 n=9 | 202±75** n=10 | 21 | 2490±498 n=9 | 1031±354* n=10 | 41 | 7094±1251 n=8 | 2660±1055* n=9 | 37 |
| Lewis lung carcinoma | 321±48 n=11 | 98±21** n=10 | 31 | 2601±269 n=11 | 755±163** n=10 | 29 | 6721±549 n=11 | 1720±402*** | 28 |
| Melanoma B-16 | 527±204 n=8 | 87±32 n=8 | 17 | 3788±1113 n=8 | 1228±334* n=8 | 32 | 8247±1338 n=8 | 4151±904* n=8 | 50 |
| Sarcoma S-180 | 1128±178 n=10 | 208±79** n=10 | 18 | 2177±327 n=10 | 654±233** n=9 | 30 | 2132±406 n=10 | 841±318* n=9 | 39 |

*$p<0.05$; **$p<0.01$; ***$p<0.001$.

We studied the effect of MP-2 in detail upon the development of sarcomá S-180. This tumor was capable of spontaneous regression that began 3 weeks after implantation. However, despite regression of the tumor, the mice died during the time period of 20th to 40th day after the tumor transplantation due to metastases that were induced before the tumor regressed.

The data presented in Table 32 show that MP-2 injection increased the life span of tumor-bearing mice—the average life span of mice after MP-2 injection increased by 45%. It should be noted that seven out of a cohort of 20 animals in the experimental group had a life span longer than 50 days and two mice were alive up to the 90th day (when they were sacrificed). In the control group the maximum life span was only 43 days (three mice out of 20).

Table 32. MP-2 effect on the life span of animals and tumor state
in mice with transplanted sarcoma S-180.

| Animal groups | Life span (days) | Tumor state | | | |
|---|---|---|---|---|---|
| | | Growth | Stabilization | Partial regression | Total regression |
| Control (n=20) | 35±1.3 (100%) | 17 (85%) | 2 (10%) | 1 (5%) | 0 (0%0 |
| MP-2 treated mice (n=20) | 45.5±1.9 (145%) p<0.05 | 11 (55%) | 2 (10%) | 5 (25%0 | 2 (10%) |

When considering the data on tumor regression, one can see that MP-2 injection promoted this process. In the experimental group the percentage of mice with actively growing tumor was lower when compared to the control, while the number of animals with partial or full tumor regression was higher. The number of mice with stabilized tumors was the same in the experimental and control groups. Thus the increase in the life span of tumor-bearing animals caused by MP-2 injection is apparently connected with partial or full tumor regression, but not with its stabilization.

The antitumor effect of MP-2 was displayed not only in implanted mouse tumors but also in spontaneous mammary adenocarcinoma. BLDR mice are predisposed to this tumor: it develops in 100% of female mice within the first year after confinement (Moiseeva *et al.*, 1991). Experimental and control groups were composed of mice having tumors of the same size (13–15 cm). The experimental mice were injected with MP-2 twice at a dose of 1 mg/kg with a 96 h interval between the injections, whereas the control mice were injected with physiologic saline at the same times.

Table 33. MP-2 effect on spontaneous growth of breast adenocarcinoma
in BLRB mice .

| Animal groups | Mice with inhibited tumor growth* after MP-2 injections at various terms | | |
|---|---|---|---|
| | $4^{th}$ day | $10^{th}$ day | $15^{th}$ day |
| Control | 5 of 13 (38%) | 4 of 13 (31%) | 3 of 12 (25%) |
| Experimental | 7 of 12 (58%) | 6 of 12 (50%) | 7 of 12 (58%) |

*As an index of inhibition of tumor growth, we took a lack of increase in tumor size in an individual animal during the period from the last measurement, that is from the first to the fourth day, from the fourth to the tenth day, and from the tenth to the 15th day.

One can see from Table 33 that the number of females with inhibited tumor growth in the group receiving MP-2 was higher than in the control group at all three observation periods. The antitumor effect of MP-2 was also revealed by a longer life span of these animals. The mean life span of mice in the control group was $42.6\pm4.9$ days (n=25) while in the group treated with MP-2 it was $63.4\pm6.6$ days (n=25). So the life span of mice increased 1.5-fold under the influence of MP-2. The differences in mean life span in these groups were statistically significant ($p<0.05$).

### 3.3. Myelopeptide 2 Antitumor Effect in vivo is Mediated by the T system of Immunity

The data presented in Section 3.1 show that the immunoregulatory activity of MP-2 is displayed by its ability to recover T lymphocyte function that is inhibited by tumor toxins. This effect of MP-2 is accompanied by CD3 and CD4 phenotype correction. It is known that CD3 and CD4 antigens play a key role in the activation and functioning of T lymphocytes, and in the development and spreading of tumors associated with the reduction of CD3+ and CD4+ T cell number (Renner and Pfreundschuh, 1995; Weil-Hillman *et al.*, 1991).

Thus we suppose that the MP-2 antitumor action *in vivo* is due to its restoring effect upon reduced T lymphocyte function in the tumor-bearing body, that is, the effect is realized through the T system of immunity. To confirm this assumption we conducted experiments on thymus-deprived nude mice with transplanted melanoma B-16.

Table 34. Change in the tumor size after melanoma B-16 transplantation to nude mice treated with MP-2.

| No of experiment | Animal groups | Days after tumor transplantation and tumor size | | | | | | | |
|---|---|---|---|---|---|---|---|---|---|
| | | 10 | | 13 | | 17 | | 20 | |
| | | Size, mm$^3$ | % of the control | Size, mm$^3$ | % of the control | Size, mm$^3$ | % of the control | Size, mm$^3$ | % of the control |
| 1 | Control | 2300±377 n=8 | 100 | 4992±647 n=8 | 100 | 8261±1813 n=7 | 100 | 12056±3438 n=5 | 100 |
| | MP-2 | 1853±135 n=9 | 80.5 | 4592±394 n=8 | 93 | 7783±431 n=7 | 94 | 12341±1097 n=6 | 101.5 |
| 2 | Control | 273±33 n=12 | 100 | 1799±202 n=11 | 100 | 6800±736 n=11 | 100 | 7570±907 n=11 | 100 |
| | MP-2 | 410±34 n=12 | 150 | 1701±160 n=12 | 94 | 6903±793 n-12 | 101 | 8725±855 n=11 | 115 |

In the first experiment, nude mice three days after tumor implantation were twice injected with MP-2 at a dose of 1 mg/kg. The interval between the injections was 96 h. In the second experiment the dose of implanted tumor cells was reduced twofold, and MP-2 treatment began one week after tumor transplantation. The control groups were injected with physiologic saline instead of MP-2. The tumor size was measured on the 10th, 13th, 17th, and 20th day after tumor implantation. The results of the experiments are shown in Table 34.

The tumor size in test and control animal groups did not differ at any time interval (p>0.05). Thus, the T system of immunity is essential for the antitumor effect of MP-2, and T lymphocytes are apparently the target cells for the action of this peptide. Thymus-deprived nude mice have no T lymphocytes, so MP-2 cannot display its antitumor effect in these mice.

The key role of the T system of immunity in MP-2 antitumor action is indirectly substantiated by experimental data on the modifying action of MP-2 on the antitumor effect of cytokine IL-2 and the cytostatic drug cisplatinum (Rosenberg *et al.*, 1969). The experiments on combined treatment with IL-2 and MP-2 were carried out on the transplantable mouse melanoma B-16. The animals were divided into four groups. The mice in the first group were injected with physiologic saline on the third day after tumor transplantation (control); the animals in the second group were injected with MP-2 at a dose of 0.5 mg/kg daily, for seven days; the mice in the third group were injected with IL-2 at a dose of 1000 IE/mouse, seven injections daily; in the fourth group the mice were treated with both IL-2 and MP-2 at the same doses and the same time intervals. The tumor size was measured on days 11, 17, and 24 after melanoma B-16 transplantation. The data presented in Table 35 show that the combined administration of MP-2 and IL-2 to mice does not augment their inhibitory effect on tumor growth as compared to the separate administration of these agents. Moreover, in some cases the tumor size displayed enhanced growth under combined treatment with IL-2 and MP-2 in comparison to treatment with IL-2 alone.

We obtained different results when we used combined treatment of mice with MP-2 and cisplatinum. The experiments were carried out on mice with implanted lympholeukemia P-388. MP-2 at a dose of 2 mg/kg was injected three times. The interval between the injections was 96 h (second group). Cisplatinum was injected once at a dose of 4 mg/kg (third group). The animals in the fourth group were treated with MP-2 and cisplatinum at the same doses and the same time intervals. The tumor size was measured on days 11, 14, 18 after tumor transplantation.

Table 35. Tumor growth in mice with melanoma B-16 treated with MP-2 and IL-2.

| Animal groups | Days after tumor transplantation and tumor size | | | | | |
|---|---|---|---|---|---|---|
| | 11 | | 14 | | 18 | |
| | $mm^3$ | % | $mm^3$ | % | $mm^3$ | % |
| Control | 1090±360 n=7 | 100 | 3160±590 n=7 | 100 | 6299±1028 n=5 | 100 |
| MP-2 | 569±190 n=7 | 52 | 1864±410 n=7 | 59 | 4397±870 n=5 | 69 |
| IL-2 | 44507±121 n=7 | 42 | 1318±310 n=7 | 42 | 3240±780 n=6 | 51 |
| MP-2 + IL-2 | 610±210 n=7 | 56 | 2430±480 n=6 | 77 | 3505±610 n=6 | 56 |

Table 36 shows that the injections of MP-2 and cisplatinum have a greater inhibitory effect on tumor growth as compared to the separate injections of these agents. The augmentation of the antitumor effect under combined cisplatinum and MP-2 treatment was maximal when the dose of cisplatinum was very low. Figure 42 shows the ratio between the tumor size in animals before treatment and after the injections of MP-2 and cisplatinum, that is the degree of tumor growth during eight days of observation (Vt/Vo). It is evident that larger doses of cisplatinum (8 or 4 mg/kg) caused significant inhibition of tumor growth; its combination with MP-2 yielded a threefold augmentation of the effect.

Small doses of cisplatinum (five injections at a dose of 0.1 µg/kg, in all 0.5 µg/kg) caused a slight inhibition of tumor growth (40% only), but when combined with MP-2 it resulted in a 85% inhibition of tumor growth. This effect exceeded that of cisplatinum at a dose of 4 µg/kg. Thus MP-2 allowed the effective dose of the toxic cytostatic drug to be decreased eightfold.

Table 36. Tumor growth in mice with lympholeukemia P-388
treated with MP-2 and cisplatinum.

| Animal groups | Days after tumor transplantation and tumor size | | | | | |
|---|---|---|---|---|---|---|
| | 11 | | 14 | | 18 | |
| | $mm^3$ | % | $mm^3$ | % | $mm^3$ | % |
| Control | 311±92 n=7 | 100 | 2111±586 n=7 | 100 | 5193±1009 n=6 | 100 |
| MP-2 | 141±43 n=7 | 45 | 1353±432 n=7 | 64 | 3888±1233 n=7 | 75 |
| Cisplatinum | 60±23 n=7 | 19 | 262±61 n=7 | 12 | 863±221 n=7 | 16 |
| MP-2 + cisplatinum | 15±7 n=7 | 5 | 99±25 n=6 | 5 | 467±88 n=7 | 9 |

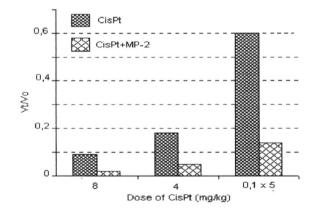

Fig. 42. Efficacy of MP-2 and cisplatinum combined application at various cisplatinum doses (on the eighth day after implantation of lympholeukemia P-388 to mice).

The data obtained substantiate the fact that the antitumor effect of MP-2 is realized via the T system of immunity. Since T lymphocytes are the target cells for both MP-2 and IL-2, their effects are not additive under combined injections of these agents. As to the combined treatment with MP-2 and cisplatinum, the cytostatic drug directly acts upon tumor cells on the background of the reduced activity of the immune system impaired by tumor toxins. Restoring the T lymphocyte activity in a tumor-bearing body, MP-2 recruits the immune system to resist the tumor, thus enhancing the direct antitumor effect of cisplatinum on the tumor.

We believe that MP-2 is a promising compound to be used in antitumor therapy. It corrects disorders in the T system of immunity by restoring the phenotypic shifts on the membrane of T lymphocytes induced by tumor cell products. Thus MP-2 apparently participates in antitumor immune defense of a tumor-bearing organism. This low molecular weight peptide is nontoxic, which is confirmed by the absence of any side effects of the preparation Myelopid which includes MP-2 as a component. Our knowledge on its amino acid sequence enables one to synthesize its analogs and to choose the form which is optimal for clinical application. Thus, the peptide Form–Leu–Val–Val–Tyr–Pro–Trp termed Bivalfor (Strelkov *et al.*, 1995) is resistant to aminopeptidases, that is, it can significantly prolong the curative effect when used in complex anticancer therapy.

As to the physiological significance of MP-2 and its role in restoring the normal functioning of the organism, we assume that this immunoregulatory peptide is anticarcinogenic by preventing tumor growth under various circumstances. The immunological surveillance performed by the immune system eliminates all foreign substances including tumor cells. One of the key processes in tumor development is suppression of the immune system of the host and disorders in T lymphocyte function, in particular. MP-2 corrects these disorders, thus promoting the retention of antitumor immunity. This is apparently what the role of MP-2 is in the mechanisms that prevent the appearance of malignant tumors.

## 4. Conformational Study of Hexapeptides MP-1 and MP-2

It was shown earlier that MP-1 and MP-2 have different biological activities — MP-1 displays an immunocorrecting effect in immunodeficient states of various etiology, while MP-2 depresses the growth of various types of solid tumors in mice. Both peptides function by affecting T cell immunity.

When one analyzes the amino acid sequences of MP-1 and MP-2, one's attention is attracted by some similarity in their primary structures: both peptides are composed of six amino acid residues with a proline residue in the fifth position, that restricts the mobility of the C-terminal fragment of the peptide chain; both peptides have mainly hydrophobic amino acid residues in their composition. Hence one could expect some similarity in the properties of these peptides. However the different biological activities of MP-1 and MP-2, and the absence of competitive interaction in binding with specific receptors, suggests that these peptides have different spatial structure and different "biologically active" conformations.

To establish the structure–function correlation with biological activities of MP-1 and MP-2 and to obtain data on the three-dimensional structure of the molecules in solution, we studied these peptides by means of physico-chemical and mathematical methods.

The first step of our investigation was to reveal the minimal amino acid sequence that retains biological activity of the respective peptide. For this reason, we synthesized MP-1 and MP-2 analogs that lack either an N-terminal (Leu–Gly–Phe–Pro–Thr and Val–Val–Tyr–Pro–Trp) or a C-terminal (Phe–Leu–Gly–Phe–Pro and Leu–Val–Val–Tyr–Pro) amino acid end. The results of biological testing in appropriate test systems showed that shortening of the peptide chain of either of the peptides by a single amino acid residue caused a complete loss of their specific activity—that is hexapeptides MP-1 and MP-2 are minimal biologically active structures.

As a rule, small peptides have no fixed spatial structure in solution; they exist as a set of molecules with various three-dimensional structures. The relative quantity of conformers is determined by the physico-chemical properties of the solvent. The biologically active peptide conformation, which occurs through its interaction with the receptor, is not always included in this set. The peptide–receptor interaction can cause a conformational rearrangement of the peptide backbone thus providing efficient interaction between the active peptide site and the ligand-binding receptor site. It is an extremely crucial task to reveal active sites and biologically active conformations of regulatory peptides, since it allows one to approach directed design and synthesis of their analogs with predetermined properties.

Determination of biologically active conformations of linear peptides is a very complicated problem that can be solved by means of various spectral methods. The most informative and widely used are the spectroscopic methods capable of providing information on the molecular structure of peptides in solution—circular dichroism (CD), Fourier transform infrared (FTIR) and nuclear magnetic resonance (NMR) spectroscopy. Mathematical methods (theoretical conformational analysis) capable of assaying the geometrical parameters of a molecule can make a valuable contribution to the study of spatial structure of small peptides. However, it should be noted that the results of theoretical calculations have to be confirmed by experiments.

The three-dimensional structure of MP-1 and MP-2 was studied by means of CD and FTIR methods and theoretical conformational analysis. CD methodology is particularly well suited to monitor changes in conformation induced by external factors, such as the nature of the solvent. FTIR spectroscopy became a valuable technique to investigate secondary structure. These methods allow one to identify secondary structure elements in the peptide molecule, such as α-helix or β-turn which play an important role in immunological processes (Dyson et al., 1985; Hollosi et al., 1990).

Theoretical assay of possible low energy conformers for MP-1 and MP-2 was carried out using a semiemperical method developed by Popov (Akhmedov et al., 1986). The values of the torsion angles of backbone and side chains that were obtained were used for a graphic presentation of the spatial structures of MP-1 and MP-2 (Fig. 43).

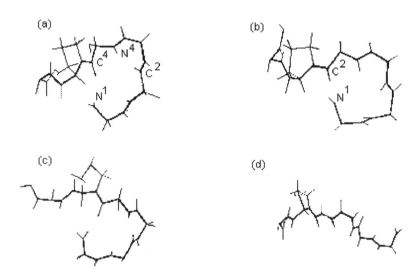

Fig. 43. Proposed low energy conformations (U total<4 kKal/M) of MP-1 (a, c) and MP-2 (b, d).

Our evaluations demonstrated that folded forms are energetically advantageous for MP-1 (Fig. 43a, c), while for MP-2 both folded and linear structures are well suited (Fg. 43b, d). Analysis of these three-dimensional structures points to the possibility of stabilization of the folded structures of both peptides by 1-4 intramolecular H-bonds between the $N^1$–H group of the first amino acid residue and the $C^4$=O group of the fourth residue (for MP-1 this dimension is 2.53 Å and for MP-2 it is 3.0 Å, Fig. 43a, b). The folded conformation of MP-1 can also be stabilized by an additional H-bond between $C^2$=O and H–$N^4$ (bond length is 3.2 Å, Fig. 43a).

The possibility of H-bonds that stabilize the folded MP-1 and MP-2 structures in solution was analyzed by the FTIR method. Infrared spectra of both peptides are shown in Fig. 44. In the region of the amide A (N–Hbending) modes there are strong bands at 3353 and 3288 cm$^{-1}$ (MP-1 and MP-2, respectively) attributed to hydrogen bonded NH-groups. In the region of the amide I (C=O-stretching) modes one can find bands at 1636 and 1646 cm$^{-1}$ (MP-1 and MP-2, respectively) attributed to H-bonded amide C=O groups. According to literature data, the frequency values that we obtained are characteristic for intramolecular H-bonds of 1-4 type stabilizing β-bends (Hollosi *et al.*, 1994; Toniolo *et al.*, 1995).

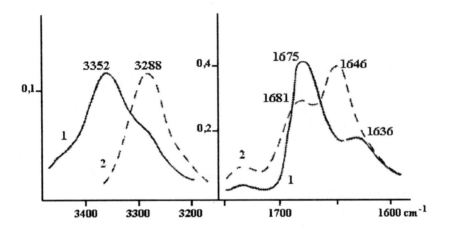

Fig. 44. FTIR spectra of MP-1 (1) in tetrahydrofuran and MP-2 in trifluoroethanol (2) solutions in the region of C=O stretching and NH-bending bands.

Thus the results of spectral studies on peptides in solution confirm the conclusions of the theoretical conformational analysis of possible stabilization of the spatial structures of peptides of interest by intramolecular hydrogen bonds.

As mentioned above, the interaction between the peptide and its receptor can cause a conformational change in the peptide chain—that is the peptide acquires a biologically active conformation under the microenvironment at the receptor. Therefore it was interesting to study the effect of the environment on the conformational state of the peptides. Conformational flexibility of the peptide chains of MP-1 and MP-2 in solution was assessed by means of CD spectroscopy. Since the receptor biophase can have different polarity, various solvents were used to analyze the peptides (water, trifluoroethanol, dioxan and tetrahydrofuran). Such a broad selection can be justified in terms of different physico-chemical properties of the solvents, as well as of different effects on the three-dimension peptide structure. Dioxan and tetrahydrofuran are organic solvents with hydrogen-bond acceptor ability; trifluoroethanol is an organic solvent with hydrogen-bond donor ability; while water is a universal solvent with hydrogen-bond donor/acceptor ability. CD spectra for both peptides are shown in Fig. 45a, b).

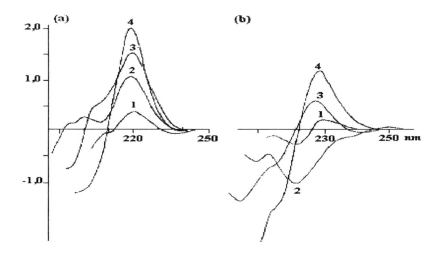

Fig. 45. CD spectra of MP-1 (a) and MP-2 (b) in dioxan (1), tetrahydrofuran (2), trifluoroethanol (3) and water (4).

The type of solvent results in CD spectrum changes. This points to a conformational flexibility of peptide chains and an environmental dependence of peptide spatial structure. CD spectra were analyzed taking into account the data on small model peptides of similar structure obtained by Hollosi and co-workers (Hollosi *et al.*, 1994). They studied the correlation between the form of CD curves and the presence of secondary structure elements. Comparison of our CD curves with literature data suggests the presence of ordered structures with β-bends in MP-1 (in trifluoroethanol) and MP-2 (in tetrahydrofuran). A mirror character of CD curves of MP-1 and MP-2 in these solvents points to a different spatial orientation of carbonyl groups in amide bonds and β-bends of different types.

Hence this study of MP-1 and MP-2 by spectral methods showed that under certain conditions these endogenous peptides can have ordered structures with β-bends stabilized by H-bonds. Experimental data presented in this chapter show that in spite of a similarity in the primary structures of MP-1 and MP-2, their conformational possibilities and spatial structure differ greatly. This can explain the different biological activities of these peptides and the absence of any competition in binding with specific receptors.

## 5. MP-3 is a Myelopeptide Stimulating Macrophage Phagocytosis

As was already mentioned earlier, we searched for activities of every newly isolated, identified and synthesized MP based on our knowledge of the whole spectrum of biological activities previously revealed in their mixture and we therefore used systems in which these effects could be detected. One of these test systems was the assay of mouse peritoneal macrophage phagocytic activity in the NBT test (Rook *et al.*, 1985). The MPs were added to mouse peritoneal macrophages cultivated with opsonized SRBC. The level of SRBC phagocytosis by macrophages was assessed by the change in optical density of the solution containing nitroblue tetrazolium reduced to formazan by oxygen released during phagocytosis. MP-1, MP-2 or MP-3 were studied at a wide range of concentrations ($1 \times 10^{-5} - 1 \times 10^{-19}$ g/ml). Data presented in Fig. 46 show that among all MPs tested only MP-3 increased the macrophage phagocytic activity.

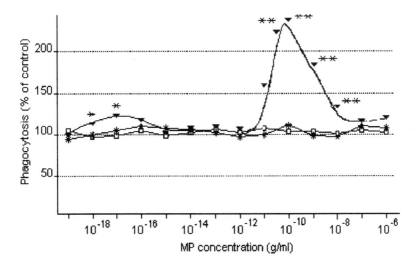

Fig. 46. Stimulating effect of MP-3 on SRBC phagocytosis by peritoneal mouse macrophages. —*— MP-1; —□— MP-2; —▼— MP-3; * p<0.05; ** p<0.01.

The MP-3 effect was dose-dependent with two stimulation peaks. In the concentration range of $0.5 \times 10^{-7} - 0.5 \times 10^{-9}$ g/ml it increased phagocytosis up to 163–232% (p<0.05), while in the second concentration range ($1 \times 10^{-15} - 1 \times 10^{-17}$ g/ml) the stimulation reached 115-124% (p<0.05). Neither MP-1 nor MP-2 displayed such effect at any concentration used.

It was also shown that MP-3 slightly changed the macrophage adhesive ability. MP-1 and MP-2 produced no effect on macrophage adhesion at all, while MP-3 enhanced adhesion by 17–25% at doses of $1 \times 10^{-8} - 1 \times 10^{-12}$ g/ml (Fig. 47).

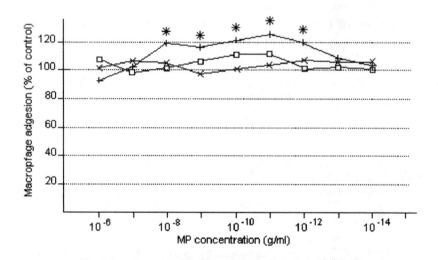

Fig. 47. MPs effect on the adhesion of mouse peritoneal macrophages. —□— MP-1; —×— MP-2; —+— MP-3. * p<0.05.

Having discovered the ability of MP-3 to enhance macrophage phagocytosis, we investigated its possible protective effect in animals infected with pathogenic microorganisms (Petrov *et al.*, 1997-b). (CBA×C57BL)F$_1$ mice were injected intraperitoneally with MP-3 at doses of $1 \times 10^{-6}$ or $0.5 \times 10^{-4}$ g/mouse. 24 h later the mice were infected with pathogenic *Salmonella typhimurium* strain 415 at various doses—$10^2$, $10^3$,

$10^4$, and $10^5$ bacterial cells/mouse. Every group consisted of 10 mice, and the observation period was for 21 days. Figure 48 shows data on animal survival on the 21st day in experimental and control groups.

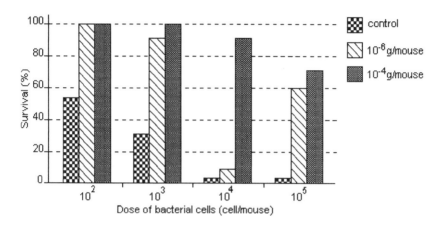

Fig. 48. Protective effect of MP-3 when injected to (CBA×C57BL)F$_1$ mice 24 h before infection with *Salmonella typhimurium* 415.

Infection with lethal doses of *S. typhimurium* ($10^4$ and $10^5$ cells/mouse) caused 100% animal death on the 21st day in the control group, while in the experimental groups injected with MP-3, 60–90% of the animals survived. The infection of mice with lower doses of bacterial cells ($10^2$ and $10^3$ cells/mouse) provoked a 50–70% animal death. In the experimental animals treated with MP-3, we observed a 100% survival. It should be noted that MP-3 influenced the dynamics of death of infected animals as well. The first deaths in the control groups infected with lethal doses of *S. typhimurium* ($10^4$ and $10^5$ cells/mouse) were registered on the fourth to fifth day, the rest of the mice dying on the 11th–20th day. In the groups injected with MP-3, the mice began to die on the 10th–11th day, and by the 21st day only 10–40% experimental mice had died. Lower doses of bacterial cells ($10^2$–$10^3$ cells/mouse) caused lethality starting from the 8th–10th day, and by the 21st day 50-70% of control animals had died, while in four experimental groups treated with MP-3 only one of the mice had died by the 20th day.

Based on the experimental data presented, one can conclude that hexapeptide MP-3 has an immunostimulating effect on the macrophage link of immunity. It stimulates the capture and elimination of foreign agents at the first stage of immune defense, thus reducing the number of pathogenic microorganisms in the body. This ability of MP-3 can explain its protective action when injected to mice 24 h prior to *S. typhimurium* infection (Fig. 48). Since MP-3, like the other individual peptides, is a component of the Myelopid preparation, it is evidently responsible for the antibacterial effect of Myelopid.

Further studies of MP-3 action on other biological properties of the macrophage — this multifunctional immunocompetent cell—will allow us to study the mechanism of the directed regulatory action of MP-3 on the macrophage link of immunity in detail.

## 6. Myelopeptide 4 is a Cell Differentiation Factor

Based on the information pooled from the first part of this book, there are substances in the mixture of MPs that are capable of influencing cell differentiation. Treatment eith MPs of $W^v/W^v$ mice suffering from the syndrome of inherited anemia restored hemopoiesis in the defective bone marrow (Petrov *et al.*, 1990-b); MPs altered *in vitro* the metabolism of bone marrow cells from patients with acute myeloleukemia (Strelkov *et al.*, 1989), thus displaying the properties of a cell differentiation factor. MPs also influenced the differentiation of T lymphocyte precursors in the bone marrow. Three hour incubation of mouse bone marrow cells with MPs resulted in a nonreversible decrease in the expression of Sc-1 antigen and an increase in the expression of Thy-1 antigen (Petrov *et al.*, 1989). Finally, the preparation Myelopid, having MPs as its acting basis, improved the immune and hematological status in acute leukemia patients. The patients treated with Myelopid displayed inhibition of leukemia tumor and normalization of the cellular composition of the bone marrow, thus achieving rapid remission (Stepanenko *et al.*, 1989).

Taking into consideration all these data, we tested our individual MPs in a system that allows one to measure cell differentiation using simultaneous assessment of DNA and protein synthesis by means of double radioactive labels ($[^3H]$ thymidine and $[^{14}C]$ glycine, respectively).

As mentioned above, a decrease in DNA synthesis (cell proliferation) and a simultaneous increase in protein synthesis (cell maturation) is the basis for assaying cell differentiation. The human myelomonoblastosis cell line HL-60 was used as a model. The HL-60 cell line was maintained in standard medium (RPMI 1614) with addition of 15% fetal calf serum, 20 mM HEPES buffer, 2 mM L-glutamine and antibiotics. The cells obtained at the logarithmic growth stage (on the third day) were washed and cultivated with MP-4 for three days; then the cells were washed once again and cultivated without MP-4 for another three days. MP-4 was used in a wide concentration range (0.01–1.00 µg/ml). As positive controls for the effect of MP-4 on cell differentiation we used two well characterized inducers of HL-60 cell differentiation: namely phorbol ester (PMA) and maturation inducer (MI) — a lymphokine contained in the supernatant from mitogen-activated T lymphocytes.

The labels ([$^3$H] thymidine and [$^{14}$C] glycine) were added to the culture 4 h prior to the termination of cultivation, that is at the end of the sixth day.

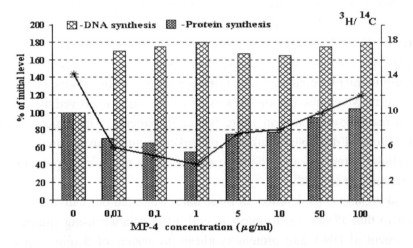

Fig. 49. MP-4 effect on the ratio of DNA and protein synthesis in the human myeloblastic leukemia cell line HL-60. Columns show the synthesis as percentage of the initial level (left scale). The curve shows the change in the ratio of [$^3$H]/[$^{14}$C] under various MP-4 concentrations (right scale).

HL-60 cell morphology was determined on the eighth day of cultivation. Figure 49 presents the results of the simultaneous assay of MP-4 action on syntheis of chromosomal DNA and total protein synthesis in human leukemia myelomonoblasts (HL-60). To the left on the y-scale, the value of either DNA or protein synthesis is expressed as a percentage of the initial value (100%). These values are shown as columns. To the right on the y-scale, the $[^3H]$–DNA/$[^{14}C]$–protein ratio is shown, that is the ratio of proliferation and maturation intensity. The changes in these ratios are presented in the figure as a curve. One can see that MP-4 has a pronounced effect on the integral indices of HL-60 cell metabolism — decreasing DNA synthesis and simultaneously increasing the protein synthesis. These metabolic alterations, that indicate HL-60 cell transition from proliferation to differentiation (maturation), depend on the MP-4 concentration in the culture. The optimal differentiation effect of MP-4 was observed at a peptide concentration of 1 µg/ml — DNA synthesis decreased by 40% from the initial value while protein synthesis reached 200%. The $[^3H]/[^{14}C]$ ratio is the lowest at this MP-4 concentration.

Morphologically it was shown that at an MP-4 concentration of 1 µg/ml more than half of the cells in the HL-60 culture are represented by monocytes/macrophages, that is by differentiated forms. It should be noted that the differentiating activity is a characteristic of MP-4. Testing of other MPs in this model did not reveal similar results. For example, MP-3 has practically no effect on the metabolism of leukemia myelomonoblasts either at high (50–100 µg/ml) or at low (1–5 µg/ml) concentrations; the indices of DNA and protein synthesis in the presence of MP-3 were very close to those in the control HL-60 cultures.

Figure 50 shows the results of a morphological assay of the ability of MP-4 to induce terminal differentiation of blasts of HL-60 as compared to the routine differentiation factors of these leukemia cells. HL-60 cells can be differentiated either into the monocyte or the granulocyte pathway, depending on the differentiation inducer. Phorbol ester, or lymphokine contained in the conditioned medium of activated human T lymphocytes (maturation inducer, MI), provokes the transformation of blasts in the initial HL-60 population into mature monocytes–macrophages. MP-4 likewise induced differentiation of myelomonoblasts along the monocyte pathway. The data presented in Fig. 50 show that of all differentiation factors tested, PMA displayed the highest activity: when added to HL-60 cells it caused the appearance of 80% mature, monocytes–macrophages. A somewhat smaller

activity under the same experimental conditions was shown by T cell
cytokine MI (up to 60% initial HL-60 population were monocytes–
macrophages). The ability of MP-4 to induce the maturation of myeloid HL-
60 blasts to mature monocytes/macrophages was comparable to that of MI
(more than 50% cells become monocytes/macrophages as the result of MP-4
action). The addition of MP-3 to HL-60 cell line at concentrations of 1–
100 µg/ml did not cause any alterations in their morphology (Fig. 50).

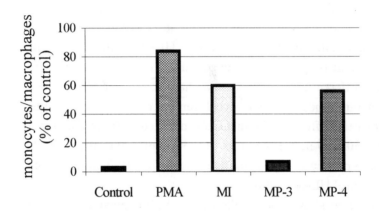

Fig. 50. MP-4 ability to induce differentiation of myeloid HL-60 cells to
monocytes/macrophages as compared to other agents.

The effect of MP-4 on the synthesis of chromosomal DNA and total
protein synthesis was also examined in another myeloid leukemia K-562 cell
line and compared to that of DMSO (differentiation inducer for erythroid
pathway). It is known that K-562 cells (erythroleukemia) respond to low
concentrations of DMSO by a decrease in their growth and start of
hemoglobin synthesis (Singer *et al.*, 1974). The effect of MP-4 on the
growth of K-562 cell was more effective then that of DMSO (1.5%).

The growth of K-562 cells was also more sensitive to the peptide as compared to that of HL-60 cells; about 50% inhibition of DNA synthesis was observed at low (0.001-0.1 μg/ml) concentrations of MP-4. Moreover, at a 100-fold lower concentration, maximal changes in $^3$H-DNA/$^{14}$C-protein ratio occurred.

Examination of the effect of DMSO on hemoglobin synthesis in K-562 cells showed that the number of hemoglobin-containing cells in cultures begin to increase by the third or fourth day after one day preincubation in 1.5% DMSO. After six days of total culture time, about 50% of K-562 cells were stained positively with the benzidine reagent if DMSO was added. MP-4 was also effective in the initiation of hemoglobin synthesis being added to K-562 cultures; over 50% of cells stained with benzidine reagent at a concentration of MP-4 of 0.001 μg/ml. There is a good correlation between the inhibition of growth and hemoglobin synthesis in K-562 cells under the effect of MP-4.

Thus the antiproliferetive effect of MP-4 is not only attributable to leukemia the HL-60 cell line. Other myeloid leukemia cells K-562 are also differentiated and mature under the influence of this peptide.

The experimental data presented allow one to regard the octapeptide of endogenous origin, namely MP-4, as a new cell differentiation factor. Antitumor therapy by induction of malignant cell differentiation ("differentiation therapy") finds a wide use in the clinic, in particular, for treatment of acute leukemia (Degos, 1990; Pui, 1995). This method is based on the ability of some agents (cytostatics, recombinant CSFs) to produce an effect on the damaged proliferation and differentiation balance in the tumor-bearing organism (Degos, 1990). They suppress the proliferation (DNA replication) of tumor cells thus inducing their differentiation; alternatively the induction of differentiation causes the suppression of cell multiplication. The effect of MP-4 on tumor cells is induction of metabolic alterations typical for normal cell differentiation. It is therefore a promising endogenous compound to use as a bioregulator in leukemia therapy.

## 7. Myelopeptides: A New Class of Endogenous Bioregulatory Peptides

Most peptides are universal bioregulators. They are produced by cells of various systems in an organism and fulfill regulatory functions specific to a given system and they participate in intersystem interactions. Thus, nervous system cells produce a group of neuropeptides, the opioid peptides, that are mediators of the nervous system; thymus cells produce a group of immunopeptides that are responsible for maturation and differentiation of T lymphocytes. The opioid peptides also have immunomodulatory properties by interacting with specific receptors on the surface of lymphoid cells. Conversely thymus peptides bind to opioid receptors on neurons (Zozulya *et al.*, 1985).

In this book it is shown that bone marrow cells produce a group of peptides with a wide spectrum of biological activities. The question arises: do the identified peptides exist *in vivo* or are they formed as the result of enzymatic proteolysis in the course of isolation? To answer this question unambiguously, additional experiments are necessary, but there are a number of indirect facts that attest to the endogenous nature of the peptides isolated:

➢ Comparison of the chromatographic profiles of the mixture of MPs from different lots of the supernatant indicate a relative constancy of peptide composition;

➢ The isolated peptides are relatively stable. We used comparatively mild conditions in our isolation procedure that are unlikely to cause chemical degradation of protein or polypeptide molecules;

➢ Several studies on the isolation and determination of the peptides from different lots of the supernatant yielded analogous results;

➢ None of the isolated peptides was revealed in the homogenates of bone marrow cells;

➢ The peptides are produced by bone marrow cells in the course of their normal vital function without any stimulation;

➢ For some pathological states (agammaglobulinemia, myeloblastic leukemia *etc.*) diminished production of MPs is characteristic;

➢ Every isolated peptide displays its own specific activity, chracteristic for this very peptide;

➢ The peptides demonstrate their activities in a dose-dependent manner *in vitro* and *in vivo*; the concentrations at which the peptides act are rather low, which is characteristic of endogenous peptide bioregulators;

➢ The existence of specific receptors on the target cells.

While the endogenous origin of the investigated MPs raises no doubts, the question of what the precursor proteins for these peptides are is still not solved.

As mentioned above, our search for homologies through the bank of known protein amino acid sequences demonstrated that the amino acid sequences of peptides MP-1 and MP-2 are homologous to the conservative fragments –(33–38)– and –(31–36)– of hemoglobin chains, respectively. Peptides MP-3, MP-4, MP-5 and MP-6 proved to have no homologous sequences in the bank; they are obviously fragments of some unknown precursor proteins. In spite of the complete homology of MP-1 and MP-2 amino acid sequences with the mentioned hemoglobin fragments, it cannot be excluded that these peptides may originate from precursor proteins other than hemoglobin. Hemoglobin fragments corresponding to the amino acid sequences of MP-1 and MP-2 are not restricted by base amino acid pairs in the protein chain, which was previously shown for precursor proteins, e.g. proopiomelanocortin. Thus they cannot be products of routine processing. We assume that these peptides are either proteolysis products of a previously unknown precursor protein or products of unspecific hemoglobin degradation.

In spite of the fact that the investigation described above is far from complete, the results obtained at this stage permit certain conclusions to be drawn. Namely those active peptides produced by bone marrow cells are apparently a new class of immunoregulatory peptides that together with thymus peptides, neuropeptides and other peptide bioregulators provide for the normal development of intercellular and intersystem interactions that occur within the organism.

The results of our studies suggest that the total number of MPs exceeds six, and there is every reason to expect that further investigations will extend the list of such peptides and expand the spectrum of their biological properties.

## Conclusion

The evidence presented on the structure, function and mechanism of action of bone marrow peptide mediators extend our knowledge of the complex network of regulatory processes that provide for normal functioning of the organism as a whole entity.

The existence of a pool of low molecular weight regulatory peptides, specifically bound to definite target cells, raises the possibility of correcting of disorders that occur in bioregulation the molecular and cellular levels.

Among the isolated and characterized MPs, we still have not revealed mediators with a neurotropic effect. This may be due to the fact that MPs with a neurotropic effect have not been isolated in a homogenous state as yet or that they lose their neurotropic effects during isolation and synthesis. Our search for MPs capable of affecting the nervous system is now in progress.

The deciphering of the structure of these bioregulatory molecules raises broad perspectives for a detailed study of their functional activity and mechanism of action.

The revelation of a directed effect of individual MPs on definitive target cells permits a new insight into the applied aspect of this problem. The medicinal preparation, Myelopid, which is formed from a natural mixture of MPs isolated from the supernatant of porcine bone marrow cell cultures, proved to be an efficient and innocuous immunocorrecting medicine which contributes to the normal functioning of the immune system in an organism weakened by disease, thus promoting a more rapid recovery. At the same time, our lack of knowledge regarding the quantitative ratio of individual MPs in different pools of the preparation, as well as the unknown composition of Myelopid, hampers its optimal application for different pathologies. The structure-function characteristic of individual MPs that make up Myelopid, and the deciphering of the mechanisms of their action raises possibilities for a scientifically based protocol for Myelopid application in specific cases.

It is important to note that the endogenous regulatory peptides we have described can be a basis for developing fundamentally new medications. Such medications may satisfy the requirements for "medicines of the future" as set out in one of the Reports of the Japan Medical Association Committee on long-term policy in new medicine development: "What clinicians seek as an ideal medication should possess hardly any or no action in a healthy person but should act only in a sick person."

There is no doubt that new MPs will be isolated and characterized in the next years. They will reveal many details of the complex and intricate problem of immunoregulation.

# References

Abbakumov, V.V., N.S. Bogomolova, R.N. Stepanenko, V.A. Ivanov, A.D. Potekhina, R.Ya. Vlasenko, V.I. Sorokina, Vestn. Akad. Med. Nauk SSSR 10 (1990) 17.

Ader, R., D.L. Felton, N. Cohen, Psychoneuroimmunology, San Diego, CA: Academia, 1991, v.II.

Adriaansen, H.J., C. Osman, J.J.M. Van Dongen, J.H.F.M. Wijdenes de Bresser, Scand. J. Immunol. 32 (1990) 687.

Akhmetov, N.A., G.A. Akhverdieva, N.M. Godjaev, E.M. Popov, Int. J. Pept. Prot. Res. 27 (1986) 95.

Ankier, S.H., Eur. J. Pharmacol. 27 (1974) 1.

Anokhin, G.N. and A.A. Yarilin, Cell Tissue Kinet. 17 (1984) 57.

Apte, R.N., S.K.Durum, J.J. Oppenheim, Immunology Letters 24 (1990) 141.

Bach, J.F., M. Dardenne, J.M. Pleau, M.A. Bach, Ann. N.Y. Acad. Sci. 249 (1975) 186.

Bandlow, C.E., M. Meyer, D. Lindholm, J. Cell. Biol. 111 (1990) 1701.

Berenbaum, M.C., Pharmaceutical J. 203 (1969) 671.

Bittiau, A., J. Van Damme, J. Ceuppens, in Lymphokine Receptor Interactions, eds. D. Fradeliz, J. Bertoglio (1989).

Blalock, J.E., Physiol. Rev. 69 (1989) 1.

Bogomolova, N.S., V.V. Abbakumov, R.N. Stepanenko, A.D. Potekhina, T.Ya. Pkhakadze, L.N. Vinogradovsa, Khirurgiia 2 (1993) 46.

Chiao, J.W., Z. Arlin, J.D. Lutten, J.S. Choi, K. Leung, Proc. Natl. Acad. Sci. USA 83 (1986) 3432.

Degos, L., Leukemia Res. 14 (1990) 717.

Derevyanchenko, I.G., S.N. Bykovskaya, L.A. Zakharova, S.V. Kibza, Biull. Eksp. Biol. Med. 101 (1986) 187.

Duwe, A.H. and S. H. Singhal, Adv. Exp. Med. and Biol. 6 (1976) 607.

Dyson, H.J., K.J. Gross, R.A. Houghten, J.A. Wilson, P.E. Wright, R.A.Lerner, Nature 318 (1985) 480.

Gilmore, W. and L.P. Weiner, J. Neuroimmunol. 18 (1988) 125.

Goldstein, A., T.L. Low, G.B. Thurman, and M.M. Zatz, in Recent Progress in Hormone Research, ed. F.O. Green (Academic Press, New York-Toronto-London, 1981).

Green, D.R. and P.M. Flood, Ann. Rev. Immunol. 1 (1983) 439.

Hollosi, M., A.Perczel, G.D. Fasman, Biopolymers 29 (1990) 1549.

Hollosi, M., Z. S. Majer, A. Magyar, S.Holly, A. Perczel, G.D. Fasman, Biopolymers 34 (1994) 177.

Isac, R., E. Mozes, M.J.Taussig, Immunogenetics 3 (1976) 409.

Ivanov, V.T., A.A. Karelin, M.M. Philippova, I.V. Nazimov, and V.Z. Pletnev, Biopolymers Pept. Sci. 43 (1997) 171.

Jahn, S., H.D. Volk, R. Grunow, S.T. Kiessig, F. Hience, E. Apostoloff, Int. J. Immunopharmac. 10 (1988) 23.

Kuo, L.M. and R.J. Robb, J. Immunol. 137 (1986) 1558.

Kuznetsova, S.F., Radiobiologiia 29 (1980) 326.

Leung, K. and J.W. Chiao, Proc. Natl. Acad. Sci. USA 82 (1985) 1209.

Lisianyi, N.I., S.A. Romodanov, A.A. Radzievskii, V.A. Rudenko, E.G. Pedachenko, Zh. Vopr. Neurokhir 3 (1991) 16.

Madden, K.S., D.L. Felten, Physiological Reviews 75 (1995) 77.

Melnik, G.V., R.N. Stepanenko, Patent Russia N 2008007 (1994).

Miescher, S., T.L. Whiteside, S. Carrel, V.J. von Fliedner, J. Immunol. 136 (1986) 1899.

Mikhailova, A.A., R.V. Petrov, R.N. Stepanenko, Dokl. Akad. Nauk 229 (1976) 247.

Mikhailova, A.A., J. Madar, V. Holan, T. Hraba, Folia Biologica 33 (1987) 50.

Mikhailova, A.A., R.V. Petrov, Sov. Med. Rev. D. Immunol. 1 (1987) 151.

Mikhailova, A.A., L.A. Strelkov, R.N. Stepanenko, E.F. Ivanushkin, Immunologiia 1 (1989) 74.

Mikhailova, A.A., S.V. Sorokin, N.A. Komponiec, Ann. Ist. Super. Sanita 27 (1991) 57.

Mikhailova, A.A., L.A. Fonina, E.A. Kirilina, S.Yu. Shanurin, S.A. Gur'yanov, A.A. Malakhov, V.A. Nesmeyanov, R.V. Petrov, Regulatory Peptides 53 (1994) 203.

Mikhailova, A.A., S.Yu. Shanurin, R.V. Petrov, Immunol. Letters 47 (1995) 199.

Miller, J.F.A.P. and G.F. Mitchell, Transplant. Rev. 1 (1969) 3.

Minami, M., J. Kuraishi, T. Jamaguchi, Biochem. Biophys. Res. Commun. 171 (1990) 832.

Mitchell, G.F., and J.F.A.P.Miller, J. Exp. Med. 128 (1968) 821.

Moiseeva, E.V., S.M. Farber, L.V. Lomova, B.V. Nikonenko, N.N. Klepikov, Laboratornye zhyvotnye 1 (1991) 24.

Nashioka, K., A. Constantopoulos, S.P. Satoh, V.A. Najjar, Biochem. Biophys. Res. Commun. 47 (1972) 172.

Nesterova, I.V., V.A. Tarakanov, A.N. Luniaka, Gematol. Transfuziol. 36 (1991) 31.

Osmanova, L.Ya., N.I. Gurarii, Zh. Mikrobiol. Epidemiol. Immunobiol. 8 (1989) 94.

Peavy, D.L. and C.W. Pierce, J. Exp. Med. 140 (1974) 356.

Petrov, R.V., Proceedings of the 12th Int. Congress on Blood Transfusion, Moscow (1969) 437.

Petrov, R.V., A.A. Mikhailova, J. Immunol. 103 (1969) 679.

Petrov, R.V., A.A. Mikhailova, Cell Immunol. 5 (1972) 393.

Petrov, R.V., A.A. Mikhailova, R.N. Stepanenko, L.A. Zakharova, Cell. Immunol. 27 (1975) 342.

Petrov, R.V., R.M. Khaitov, A.A. Batyrbekov, Dokl. Akad. Nauk Russia 226 (1976) 1446.

Petrov, R.V., R.M. Khaitov, R.I. Attaulakhanov, I.G. Sidorovitch, Dokl. Akad. Nauk 233 (1977) 745.

Petrov, R.V., A.A. Mikhailova, L.A. Zakharova, Yu.O. Sergeev, V.I. Novikov, Ann. Immunol. (Inst. Pasteur) 131D (1980) 161.

Petrov, R.V., A.A. Mikhailova, V.I. Novikov, Dokl. Akad. Nauk 258 (1981) 1252.

Petrov, R.V., P.A. Durinyan, A.M. Vasilenko, V.K. Reshetnyak, L.A. Zakharova, A.A. Mikhailova, E.O. Bragin, M.L. Kukushkin, Dokl. Akad. Nauk 265 (1982) 501.

Petrov, R.V., P.G. Deryabin, Yu.V Sergeev, N.V. Loginova, A.A. Mikhailova, G.A. Lebedeva, Zh. Mikrobiol. Epidemiol. Immunobiol. 11 (1983-a) 67.

Petrov, R.V., M.E. Vartanyan, A.A. Zozulya, A.A. Mikhailova, E.N. Patshakova, N.V. Kost, L.A. Zakharova, Biull. Eksp. Biol. Med. 5 (1983-b) 46.

Petrov, R.V., A.A. Mikhailova, L. A. Zakharova, A.M. Vasilenko, A.V. Katlinsky, Scand. J. Immunol. 24 (1986) 237.

Petrov, R.V., A.A. Mikhailova, Yu.O. Sergeev, S.V. Sorokin, EOS – J. Immunol. and Immunopharmacol. VII (1987-a) 88.

Petrov, R.V., A.A.Mikhailova, L.A.Zakharova, Ann. N.Y. Acad. Sci. 496 (1987-b) 271.

Petrov, R.V., V. S. Aprikyan, Yu. O.Sergeev, S.I.Elkina, A.A.Mikhailova, J. Microbiol. Epidemiol. and Immunobiol. 5 (1988) 62.

Petrov, R.V., S.F. Kuznetsova, A.A. Yarilin, Dokl. Akad. Nauk 5 (1989) 46.

Petrov, R.V., A.M. Borisova, A.V. Glazko, A.V. Simonova, R.N. Stepanenko, Terap. Arkh. 62 (1990-a) 81.

Petrov, R.V., A.A. Mikhailova, N.A. Kompaniec, V.M. Manko, Dokl. Akad. Nauk 310 (1990-b) 247.

Petrov, R.V., A.A. Mikhailova, Sov. Med. Rev. D. Immunol. 4 (1992) 65.

Petrov, R.V., Ann. NY Acad. Sci 685 (1993) 351.

Petrov, R.V., A.A. Mikhailova, E.A. Kirilina, Folia Biologica 40 (1994) 455.

Petrov, R.V., A.A. Mikhailova, L.A. Fonina, Biosci. Rep. 15 (1995) 1.

Petrov, R.V., A.A. Mikhailova, L.A. Fonina, Biopolymers 43 (1997-a) 139.

Petrov, R.V., A.A. Mikhailova, L.A. Fonina, Journal of Journals 1 (1997-b) 53.

Pui, C-H., New Engl. J. Med. 332 (1995) 1618.

Renner, C. and M. Pfreundschuh, Immunol. Rev. 145 (1995) 178.

Rich, R.R. and C.W. Pierce, J. Exp. Med. 137 (1973) 649.

Roitt, I., G. Torrigiani, M.F.Greaves, J. Brostoff, and J.H.L. Playfair, Lancet 2 (1969) 367.

Rook, J.A.W., J. Steel, S. Umar, H.M. Dockrell, J. Immunol. Meth. 82 (1985) 161.

Rosenberg, B., L. VanCamp, J. Trosko Nature 222 (1969) 385.

Scatchard G., Ann. NY Acad. Sci. 51 (1949) 660.

Singer, D., M. Cooper, G.M. Maniatis, P.A. Marks, and R.A. Rifkind, Proc. Natl. Acad. Sci. U.S.A. 71 (1974) 2668.

Sharp, B.M., D.T. Tsukayama, G. Gekker, W.F. Keane, P.K. Peterson, J. Pharmcol. Exp. Ther. 242 (1987) 579.

Sibinga, N.E.S. and A. Goldstein, Ann. Rev. Immunol. 6 (1988) 219.

Skryabina, E.L., M.P. Svirezheva, A.A. Mikhailova, Immunologia 4 (1987) 93.

Stepanenko, R.N., Yu. I. Skalko, A.M. Borisova, A.V. Glazko, V.I. Kuznetsov, Immunologia 5 (1989) 45.

Stepanenko, R.N., N.K. Ryazanov, O.A. Moldokulov, R.Ya. Vlasenko, Immunologia 1 (1991) 44.

Strelkov, L.A. and A. A. Mikhailova, Immunology (Russian) 6 (1989) 32.

Strelkov, L.A., R. N. Stepanenko, E.F.Ivanushkin, A. A. Mikhailova, Immunology (Russian) 1 (1989) 74.

Strelkov, L.A., and A.A. Mikhailova, Immunology 6 (1990) 32.

Strelkov, L.A., A.A. Mikhailova, L.A. Fonina, R.V. Petrov, Dokl. Akad.Nauk 338 (1994) 125.

Strelkov, L.A., A.A. Mikhailova, L.A. Fonina, S.A. Gur'yanov, R.V.Petrov, Biull. Eksp. Biol. Med. 5 (1995) 530.

Strelkov, L.A., A.A. Mikhailova, A.M. Sapozhnikov, L.A. Fonina, R.V. Petrov, Immunol. Letters 50 (1996) 143.

Taussig, M.J., Nature 248 (1974) 234.

Taussig, M.J., A.J. Munro, A.J. Campbell, C.S. David, N.A. Stainess, J. Exp. Med. 142 (1975) 694.

Toniolo, C., E. Valente, F. Formaggio, M. Grisma, G. Pilloni, C. Corvaja, A. Toffoletti, G.V. Martinez, P.M.Hanson, J. Pept. Science 1 (1995) 45.

Wiel-Hillman, G., K. Schell, D.M. Segal, J.A. Hank, J.A. Sosman, P.M. Sondel, J. Immunotherapy 10 (1991) 267.

Wybran, J., L. Schadene, J. Van Vooren, Ann. NY Acad. Sci. 496 (1987) 108.

Zakharova, L.A., R.G. Belyovskaya, O.G. Yanovskii, Biomedical Sci. 1 (1990) 143.

Zoloedov, V.I., A.M. Zemskov, A.A. Stupnitskii, T.A. Goriainova, Klin. Med. 73 (1995) 43.

Zotova, V.V., K.S. Aslanayan, R.N. Stepanenko, Patent Russia N 1822524 (1990).

Zozulya, A.A., S.Ph. Pschenichkin, M.R. Shchurin, Y.N. Khomjakov, I.A. Besvorshenko, Acta Endocrinol. (1985) 284.